Tests for Algebra and Trigonometry

Second Edition

Functions and Applications

by

Paul A. Foerster

Addison-Wesley Publishing Company
Menlo Park, California · Reading, Massachusetts · Don Mills, Ontario
Wokingham, England · Amsterdam · Sydney · Singapore · Tokyo
Madrid · Bonn · San Juan

The blackline masters in this publication are designed to be used with appropriate duplicating equipment to reproduce copies for classroom use. Addison-Wesley Publishing Company grants permission to classroom teachers to reproduce these masters.

ISBN 0-201-25288-0

15 16 17 18 V036 13 12 11

Contents

Description of Tests

Addison-Wesley's *Algebra and Trigonometry* tests are adaptations of tests developed for use in the author's own classes. The tests are appropriate for use with both the school edition (Code No. 25086) and the college edition (Code No. 23253) of Paul A. Foerster's *Algebra and Trigonometry* text. Included is a test for each chapter, plus more detailed tests to cover material up to logical breaking points within chapters. Also included are 4 two-hour cumulative tests and a three-hour final exam that correspond to the cumulative reviews and final exam in the text.

Because most instructors prefer to write their own tests, the tests provided here are intended to guide you in writing your own tests. However, they are reproducible and may be used if the need arises and if they fit your requirements.

These tests can also be used as make-up tests for students who miss the regularly scheduled ones. If a photocopier is not available, you can give the test page to the student with instructions not to write on it. For this reason answers are separate from the tests.

Another way to create make-up (or regular!) tests is to copy problems that are examples in the text. This tactic speeds review after the test. You can say, for instance, "That was the example on page 117." If you use examples from the text a time or two, students will soon learn to read the book and work the examples before a test. Also, students cannot claim "We never saw that before!" if it was a part of their assigned reading.

It is not expected that you will give your students all 49 tests. To allow for selection of topics, there is more material in the text than can be presented in a one-year course. Obviously, you will test only the topics you have presented. Also, you will often include material from the last few sections of a chapter on the chapter test rather than giving a separate test over this material. In keeping with this idea of continuing review, most of these tests also contain material from earlier in the course. It is recommended that your own test program include this feature so that students do not forget things they once knew how to do.

It is intended that students show their work on notebook paper rather than on the test sheet. Also, it is expected that that you will evaluate students' work, not simply check their answers. However, consideration has been given to ease of grading of the tests. For example, problems that require students to draw graphs are placed at the end of the test whenever possible, so that you will not have to shuffle back and forth between sheets of notebook paper and graph paper.

It is expected that students will use calculators on all tests. No consideration of using computers has been given in writing these tests. See the Supplementary Classroom Exercises for ideas on evaluation of students' work done on the computer.

You are invited to give comments to the author concerning ideas for making these tests better. Write care of Addison-Wesley Publishing Co., 2725 Sand Hill Road, Menlo Park, CA 94025.

NAME: ____________________ DATE: ____________

Test 1 | CHAPTER 1

In first-year algebra you learned about variables, expressions, equations, and inequalities. The domain of a variable is the set of numbers to which it belongs. Give an example of

1. a negative rational number that it not an integer.

2. a positive irrational number that is not transcendental.

3. an imaginary number.

4. Explain why 35.21 is a *rational* number.

5. Evaluate the expression $3x^2 - 6x + 7$ if
a. x is 5. b. x is -4.

6. Solve the equation $3x - 7 = -40$ if the domain of x is
a. {integers} b. {rational numbers} c. {counting numbers}

7. Solve and graph the inequality $|2 - 3x| > 17$.

One kind of expression is the *polynominal.*

8. Name by degree and number of terms: $7^3x^4y^5 + z^8$

9. Name by degree and number of terms: $5x^4 - 7x^3 + 11$

10. Write an example of a quadratic monomial.

11. Write an example of an expression that is *not* a polynomial.

One technique that helps evaluate expressions is simplifying them. Simplify the following:

12. $20 - 12 \div 3 \cdot 2 + 1$

13. $|14 - 31| - 8$

14. $7 - 5[4 - 3(2 + x)]$

15. $5[2x - \frac{1}{4}(20x - 3)]$

16. Multiply: $(3x - 2)(5x + 4)$

17. Factor: $x^2 - 11x - 12$

The Field Axioms provide reasons for each step in a simplification. Use variables to state

18. the cummutative axiom for multiplication.

19. the associative axiom for addition.

20. the additive inverse axiom.

21. closure under multiplication.

Name the Field Axiom that was used:

22. $x + (y + z) = x + (z + y)$

23. $x(y + z) = xy + xz$

24. State and prove the multiplication property of equality.

NAME: ______________________ DATE: __________

Test 2 | CHAPTER 2

Algebra II is a study of functions. A function tells how two variables are related to each other. An equation may specify how they are related. For Problems 1, 2, and 3, plot the graph of the function. Tell its range.

1. $y = 0.2x^3$, domain $= \{x: -3 \le x \le 3\}$

2. $y = x - 1$, domain $= \{x: -2 < x < 4\}$

3. $y = |x - 3|$, domain $= \{\text{digits}\}$

A relation may or may not be a function. For Problems 4 and 5, tell the domain, the range, and whether or not it is a function.

4.

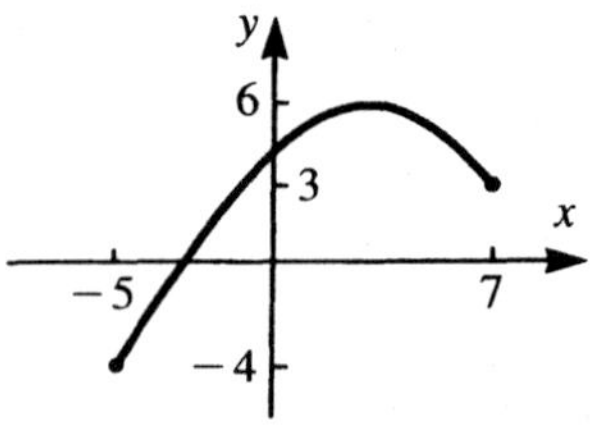

5.

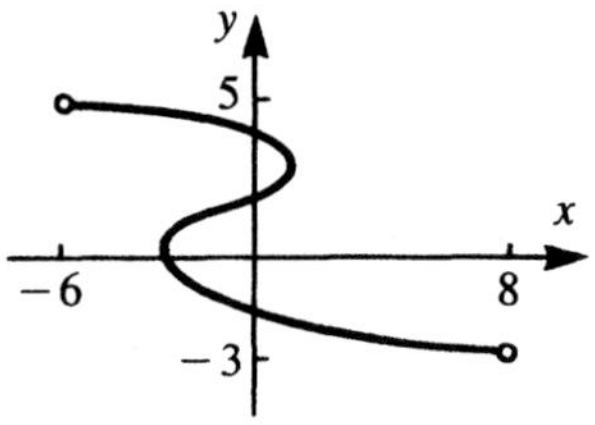

6. Plot the graph of the relation $|y| = x + 2$. Tell what the domain of x must be in order for there to be values of y. Tell whether or not the relation is a function.

Functions may relate two variables in the real world. For Problems 7, 8, and 9, sketch a reasonable graph.

7. The depth of water in a lake behind a dam depends on the number of days since they started letting water out through the flood gates.

8. The grade you could make on an algebra test is a function of the length of time you study for it.

9. The number of miles an airplane has flown on a given trip and its altitude above sea level are related.

For Problems 10 and 11, sketch the graph of the function described.

10. The function has domain $\{x: 3 < x < 8\}$ and range $\{y: -2 < y < 5\}$.

11. The graph has a vertical asymptote at $x = 4$.

It is important to be able to work old problems.

12. Plot a number-line graph: $|x + 5| \ge 2$

13. Name by degree and number of terms: $7^3x^4y^5 + z^8$

14. Evaluate: $\frac{3}{4} + \frac{1}{3}$

15. Simplify: $7 - 5[4 - 3(2 + x)]$

16. Solve: $(x + 3)(2x - 5) = 0$

NAME: ______________________________ DATE: ______________

Test 3 | CHAPTER 3, Sections 3-1 to 3-4

Algebra/Trig is a study of functions. The first kind of function you are studying is the linear function.

1. Write the general equation for a linear function.

For Problems 2 and 3, transform to slope-intercept form.

2. $3x - 7y = 42$

3. $y - 5 = -\frac{1}{2}(x - 3)$

Functions have *intercepts*.

4. Write the definition of x-intercept.

5. If $3x - 5y = 90$, find the y-intercept.

6. If $4x + 7y = 56$, find the x-intercept.

You should be able to plot the graph of a linear function *quickly* by using properties you have learned. For Problems 7 through 10, plot the graph on graph paper.

7. $y = -\frac{2}{5}x + 3$

8. $y + 6 = 2(x - 4)$

9. $y = 3$

10. $x - 4y = 12$

By using properties in the reverse direction, you should be able to find the particular equation of a linear function if you know information about the graph. For Problems 11 through 14, find the particular equation of the linear function described. Leave your answer in the most convenient form.

11. Line contains (2, 9) and (5, 4).

12. Line is parallel to the graph of $3x + 4y = -17$, y-intercept $= 5$.

13. Line contains the origin, and is perpendicular to the graph of $y = 0.25x + 11$.

14. Line is vertical, and contains (5, 8).

It is important to remember old techniques.

15. Sketch the graph of a relation that is *not* a function.

16. Write a negative even integer less than -10.

17. You drink a glass of root beer through a straw. The amount of root beer left depends on the number of sips you have taken. Sketch a reasonable graph.

NAME: ______________________________ DATE: ______________

Test 4 | CHAPTER 3, Section 3-5

Algebra/Trig is a study of functions (and relations). One of the simplest kinds of function is the linear function. On notebook paper, sketch a reasonable graph of the following linear functions:

1. Positive slope, negative y-intercept.

2. Domain: 3 to 7, inclusive; Range: 1 to 4, inclusive.

3. The amount of bismuth salicylate in a bottle of Pepto-Bismol varies linearly with the amount of Pepto-Bismol left in the bottle.

Linear functions can be used as mathematical models of relationships between two variables in the real world.

4. *Reading Problem* Phoebe Small has 35 pages of history to be read after she has been reading for 10 minutes, and 5 pages left after she has been reading for 50 minutes. Assume that the number of pages left varies linearly with the number of minutes she has been reading. Write the particular equation expressing pages in terms of minutes.

5. *Donuts Problem* The price you pay for a box of donuts varies linearly with the number of donuts in the box. For 5 donuts the price is \$1.15, and for 11 donuts it is \$2.35.

a. Write the particular equation expressing price in terms of number of donuts.
b. Predict the price of a box containing 3 donuts.
c. If a box costs \$3.15, how many donuts would you expect it to contain?
d. Sketch the graph of this linear function.
e. Tell the real-world meanings of the slope and the price-intercept.

6. *Elevator Problem* The number of feet of cable needed for an elevator depends on the number of stories in the building it serves. Suppose that $c = 20s + 35$, where c is the number of feet of elevator cable and s is the number of stories.

a. How do you know that c varies *linearly* with s?
b. How much cable is needed for a 29-story building?
c. How tall a building needs 375 feet of cable?
d. What does the slope represent in the real world?
e. What does the c-intercept equal? Why do you suppose that it is greater than zero?
f. Write a suitable domain for the linear function.
g. On graph paper, plot the graph of this function, observing the domain you wrote in part (f).

NAME: ______________________________ DATE: ______________

Test 5 | CHAPTER 3

Algebra/Trig is a study of functions (and relations). One of the simplest kinds of function is the linear function.

1. Write the general equation that defines a linear function.

The particular equation of a linear function can be found from information about its graphs. Find the particular equation.

2. Contains (3, 5) and (– 2, 7).

3. Has slope $-\frac{3}{7}$ and contains (4, – 1).

4. Has x-intercept 6 and y-intercept 9.

5. Is vertical and contains (4, 6).

6. Is horizontal and contains (3, 7).

The equation of a linear function can be transformed to other forms.

7. Transform $4x - 8y = 13$ to slope-intercept form.

8. What is the slope of the line perpendicular to the one in Problem 7 if it goes through the point (– 11, 17)?

9. $y - 5 = \frac{7}{3}(x + 4)$ is in "point-slope" form.

a. Write the coordinates of the "point."
b. Transform to $Ax + By = C$ form, where A, B, and C are integer constants.
c. Find the x-intercept.

While you are learning new things, it is important to remember old techniques.

10. Tell which axiom was used: If x and y are real numbers, then xy is a unique real number.

11. Solve the equation: $|12x - 71| = 11$

12. Multiply: $(3x + 4)(2x - 7)$

Linear functions can be used as mathematical models of relationships between two variables in the real world.

13. Suppose that a baby sitter charges \$3.50 an hour, plus \$2.00 fixed charge. Write the particular equation for this function. Then use it to predict the number of hours sat for a charge of \$21.25.

14. Linear functions can be graphed quickly. Plot the graph of the linear function in Problem 9.

NAME: ____________________ DATE: ____________

Test 6 | CHAPTER 4, Sections 4-1 to 4-3

Algebra/Trig is a study of functions. The intersection point of two linear function graphs can be found by solving a system of equations.

1. Solve by linear combination (addition-subtraction).

$$3x + 7y = -29$$
$$4x - 9y = 53$$

2. Solve by determinants (Cramer's rule).

$$5x + 2y = 8$$
$$3x - 4y = -6$$

3. On notebook paper, sketch the graphs of two linear equations that are clearly *independent*.

4. Try to solve by linear combination. Then sketch a graph showing what went wrong.

$$3x - 5y = 15$$
$$3x - 5y = 20$$

5. Try to solve by determinants. Then sketch a graph showing what went wrong.

$$8x + 12y = 36$$
$$6x + 9y = 27$$

While you are learning new things it is important to remember old techniques.

6. Write a negative decimal that is also a rational number. Then write the decimal in a form that shows *why* it is a rational number.

7. Suppose that $y^2 = x + 5$. Evaluate y when x is 11. Then explain why this relation is *not* a function.

8. On notebook paper, sketch the graph of a function with domain from 6 to 9 (*not* inclusive), and range from 1 to 7 (*not* inclusive).

9. *Computer Disk Problem* A computer store sells 10 floppy diskettes for $15, and 30 diskettes for $40. Assume that the number of dollars is a linear function of the number of diskettes. Write the particular equation expressing dollars in terms of diskettes.

10. On graph paper, plot the graph of $y = -\frac{2}{5}x - 3$.

NAME: ______________________________ DATE: ____________

Test 7 | CHAPTER 4, Sections 4-4 to 4-9

Algebra/Trig is a study of functions. In dealing with functions you must sometimes solve systems of equations with three variables.

1. Solve by augmented matrices. Express the answers as decimals.

$$\begin{aligned} 2x + 7y &= 8 \\ 5x - 3y &= 11 \end{aligned}$$

2. Solve by linear combination.

$$\begin{aligned} 3x + 2y - z &= 3 \\ x + 4y + 2z &= 19 \\ 2x + 3y - 5z &= 0 \end{aligned}$$

3. Solve the system in Problem 2 by augmented matrices.

4. Use third-order determinants (Cramer's rule) to find the value of y in Problem 2, thus showing that you get the same answer.

In mathematics it is important to remember how to solve old problems.

5. Draw a number-line graph of $|4x - 25| > 13$ if the domain of x is {natural numbers}.

When you deal with more than one function, it is convenient to use "$f(x)$" terminology.

6. Let $g(x) = 3x - 4$ and $h(x) = 2x^3$. Find $g(3)$, $h(5)$, $h(g(1))$, and $g(g(8))$.

$f(x)$ terminology is useful in some real-world problems.

7. *Swimming Problem* Adolph Finn is 130 meters from the beach. He swims toward the beach at 3.2 meters per second (m/sec). Let x be the number of seconds since he started swimming, and let $f(x)$ be the number of meters he still has to go to reach the beach.

a. Write an equation expressing $f(x)$ in terms of x.
b. Find $f(0)$ and $f(20)$.
c. Sheila Rive is only 80 meters from the beach. At the same time Adolph starts, Sheila starts swimming toward the beach at 1.7 m/sec. Let $g(x)$ be the number of meters Sheila still has to go to reach the beach. Write an equation expressing $g(x)$ in terms of x.
d. Find $g(0)$ and $g(20)$.
e. Calculate the time at which $f(x) = g(x)$.
f. Does Sheila arrive at the beach before Adolph reaches her? Explain.
g. Plot the graphs of functions f and g on the same Cartesian coordinate system.

8. Three-variable linear equations have graphs that are planes in space. Draw the graph of $3x + 2y + 5z = 30$.

NAME: ______________________ DATE: ____________

Test 8 | CHAPTER 4, Sections 4-10 and 4-11

Algebra/Trig is a study of functions. $f(x)$ terminology can be used to define functions.

1. Given $f(x) = 3x^2 - 5x + 7$, find $f(6)$ and $f(-4)$.

Systems of linear inequalities can be used to define relations.

2. Plot on graph paper the solution set of the system

$$3x - 2y \geq 12 \text{ and } y < \frac{1}{5}x + 2.$$

A system of inequalities can be used to define the domain of two independent variables in a function with three variables.

3. *Clothing Sale Problem* The Student Council decides to raise money by selling imported clothes. Let x be the number of shirts imported, and let y be the number of dresses imported. The import order must meet the following requirements.

i. The total number of shirts and dresses must be at least 10.
ii. The number of shirts must be between 5 and 27, inclusive.
iii. The number of dresses must be less than twice the number of shirts, but at least 1/6 the number of shirts.
iv. Four times the number of shirts plus three time the number of dresses must be at most 135.

Write the inequalities described, and plot the graph of the feasible region.

4. The diagram shows the feasible region for a manufacturing process in which x is the number of large machines made and y is the number of small machines made. The profit for each large machine is $1000, and the profit for each small machine is $500.

a. Write an equation for the profit in terms of x and y.
b. Shade the part of the feasible region in which the profit is at least $15,000.
c. What is the optimum number of small and large machines that produces the maximum profit? What is the maximum profit?

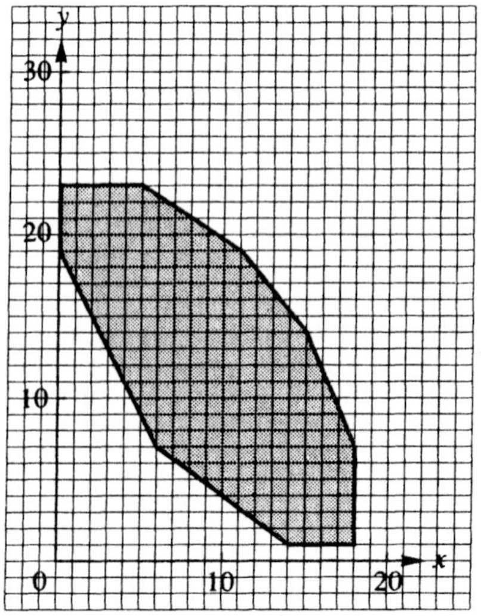

NAME: ______________________ DATE: ____________

Test 9 | CHAPTER 4

Algebra/Trig is a study of functions. Dealing with functions sometimes involves systems of equations or with two or more variables.

1. Solve by determinants.

$$\begin{aligned} 3x - 7y &= 13 \\ 2x + 9y &= -4 \end{aligned}$$

2. Solve by linear combination.

$$\begin{aligned} 5x - 4y &= -47 \\ 7x + 6y &= 27 \end{aligned}$$

3. Solve by augmented matrices.

$$\begin{aligned} 5x + 2y &= 8 \\ 3x + y &= 6 \end{aligned}$$

4. Solve.

$$\begin{aligned} 3x - 4y &= 9 \\ 6x - 8y &= 1 \end{aligned}$$

5. Solve by linear combination.

$$\begin{aligned} 2x + 3y - z &= -1 \\ -x + 5y + 3z &= -10 \\ 3x - y - 6z &= 5 \end{aligned}$$

6. Write the system in Problem 5 as an augmented matrix. Transform the matrix so that there are 0's in the lower left triangle. Use the result to find the value of z.

7. Find the value of z in Problem 5 by determinants.

8. Given $f(x) = 3x - 7$ and $g(x) = 4x + 5$, find the ordered pair where the graphs of f and g cross.

Three-variable linear equations have graphs that are planes in space.

9. Draw the graph of $3x + 4y + 2z = 12$.

Linear inequalities have graphs that are regions in the Cartesian plane.

10. Given the following requirements:

i. The sum of x and y is at most 10.

ii. y is less than twice x.

iii. y is at least 2 more than $\frac{1}{3}$ of x.

Write inequalities expressing each of these requirements. Then plot the graph of the solution of this system.

NAME: ______________________________ DATE: ____________

Test 10 | CHAPTER 5, Sections 5-1 to 5-5

Algebra/Trig is a study of functions. One kind of function you study is the quadratic function.

1. If $f(x) = 5x^2 - 8x + 13$,
  a. evaluate $f(3)$.
  b. evaluate $f(-7)$.

To evaluate x when $f(x)$ is given, you must solve quadratic equations with one variable.

2. Solve by the quadratic formula: $5x^2 - 8x - 3 = 0$

Quadratic functions have graphs that are parabolas. Sketch on notebook paper the graph of the quadratic function described.

3. Vertex at $(5, -7)$ and y-intercept 20.

4. y-intercept -3, and *no* real x-intercepts.

5. Exactly *one* x-intercept, $x = 4$.

If the equation of a quadratic function is in vertex form, you can easily sketch the graph.

6. Sketch on notebook paper the graph of $y - 8 = -3(x + 4)^2$.

If the equation is not in vertex form, you can transform it by completing the square.

7. Transform to vertex form: $y = 7x^2 + 42x - 13$

The x-intercepts of a quadratic function can be found by using the definition of intercept in connection with some of the above techniques.

8. Find the x-intercepts of $y = x^2 - 13x - 3$.

Some quadratic equations have solutions that are not real numbers.

9. Does $3x^2 + 7x + 5 = 0$ have real-number solutions? Justify your answer.

10. Given the quadratic equation $x^2 - 12x + 52 = 0$,
  a. Solve it over the set of complex numbers.
  b. Plot the two solutions in the complex plane.
  c. Write the word that describes how the two solutions are related to each other.
  d. Check one of the two solutions by substituting it into the original equation.

While you are studying quadratic functions, it is important not to forget about other techniques.

11. Find the particular equation expressing y in terms of x for the linear function containing $(-1, -2)$ and $(3, 10)$.

12. Do the squaring: $(5x + 11)^2$

13. Factor: $x^2 - 3x - 10$

NAME: ______________________ DATE: ____________

Test 11 | CHAPTER 5, Sections 5-6 and 5-7

Algebra/Trig is a study of functions. The particular equation of a quadratic function can be found from points on its graph.

1. Find the particular equation of the quadratic function containing $(-2, -11)$, $(4, 13)$, and $(6, 29)$.

2. Find the particular equation of the parabola with vertex at $(2, -5)$ and passing through $(3, 1)$.

Quadratic functions can be used as mathematical models.

3. *Pizza Problem* Pete Zarilla sells 10-inch pizzas for $4.30, 12-inch pizzas for $5.88, and 14-inch pizzas for $7.78. Assume that the price is a quadratic function of the diameter.
   a. Write the particular equation expressing price in terms of diameter.
   b. Predict the price of a 5-inch mini-pizza.
   c. If a pizza sells for $20.00, what would you expect its diameter to be?
   d. What is the significance of the fact that the price-intercept is greater than zero?

Equations of quadratic functions can be found from geometrical information.

4. An L-shaped strip x inches wide is cut from a piece of cardboard (see sketch). Write the particular equation for the area of the remaining piece of cardboard as a function of x.

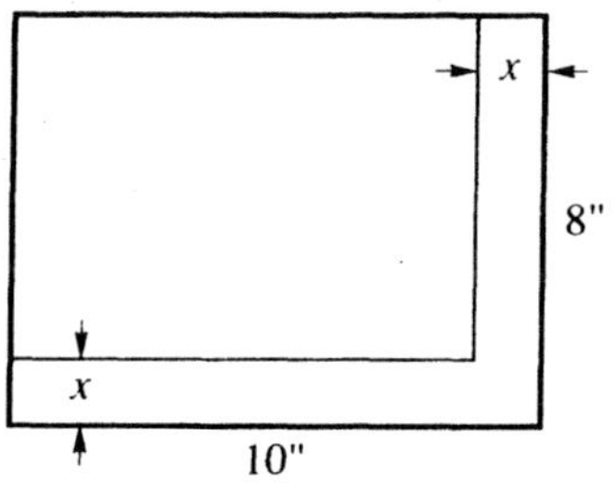

It is important to remember how to work old problems.

5. Solve the inequality $17 - 4x > 28$ and graph on a number line.

NAME: ______________________________ DATE: ____________

Test 12 | CHAPTER 5

Algebra/Trig is a study of functions. In this chapter you have learned about quadratic functions.

1. Given: $y = 3x^2 + 24x + 43$
   a. Transform to vertex form.
   b. Write the coordinates of the vertex.
   c. Find the x-intercepts.
   d. Sketch the graph, showing vertex, x- and y-intercepts, and symmetrical point.

Solving quadratic equations sometimes leads to imaginary and complex numbers.

2. Solve $7x^2 - 5x - 3 = 0$.

3. Solve $x^2 + 2x + 26 = 0$.

4. Multiply $(5 + 7i)(4 - 3i)$.

5. Given $f(x) = 11x^2 + 13x - 28$, use the discriminant to determine whether or not $f(x)$ ever equals -100 for any real values of x.

While working with quadratic functions, you must not forget about linear functions.

6. Sketch the graph of a linear function whose domain is 3 to 10, inclusive, and whose range is 1 to 4, inclusive.

Quadratic functions can be used as mathematical models.

7. *Vertical Motion Problem* Chuck Stone is standing atop a high platform. He fires a rock up into the air with his slingshot. While it is in flight, the rock's distance above the ground is a quadratic function of time. At times 1, 2, and 3 seconds after he fired it, the rock is 68, 96, and 114 meters above the ground, respectively.
   a. Write the particular equation for this function.
   b. What is the highest the rock will be above the ground?
   c. How high is the platform?
   d. The coefficient of the linear term in the equation is the initial upward velocity of the rock in meters per second. How fast was the rock going up at the instant Chuck fired it?
   e. On its way down, the rock goes into a well, splashing into the water exactly 10 seconds after Chuck fired it. How deep is the well?

NAME: ______________________________ DATE: ______________

Test 13 | CHAPTERS 1 through 5

(2 hours)

Algebra/Trig is a study of functions and relations. Write the general equation for

1. a linear function.

2. a quadratic function.

Certain axioms form the basis for algebra. Tell which axiom is illustrated.

3. $a(b + c) = ab + ac$

4. $a + (b + c) = (a + b) + c$

5. $1 \cdot (b + c) = (b + c)$

6. If $a = b + c$, then $b + c = a$.

7. Show by example why {integers} is *not* closed under division.

Axioms, properties, and definitions can be used to solve equations and inequalities.

8. Solve and graph: $|3x - 7| \leq 11$, domain = {integers}

9. Solve: $(3x - 7)(x + 4) = 0$

10. Solve: $4x^2 - 12x + 13 = 0$, domain = {complex numbers}

You must know certain names and definitions.

11. Write the definition of a relation.

12. Write a cubic binomial.

13. Write an expression that is *not* a polynomial.

14. Write the discriminant of $rx^2 + sx + t = 0$.

15. Write a negative rational number that is not an integer.

The axioms and properties can be used to transform expressions.

16. Simplify: $13 - 2 \cdot 12 \div 6 + 5$

17. Simplify: $7 - 3[p - 2(5p - 6)]$

18. Multiply: $(3x - 4)(2x + 5)$

NAME: ______________________________ DATE: ____________

(Test 13 continued)

19. Factor: $2x^2 - 11x + 12$

Graphs of familiar functions can be drawn by finding special points.

20. For $y = x^2 + 2x - 8$, find the vertex, find the y-intercept, find the x-intercepts, and sketch the graph.

21. Sketch the graph: $y - 3 = -\frac{5}{2}(x + 4)$

22. Sketch in 3 dimensions: $21x + 70y + 30z = 210$

Unfamiliar graphs can be drawn by pointwise plotting.

23. Plot $y = \sqrt{x}$, Domain: $0 \le x < 9$

24. Sketch the graph of a relation that is clearly *not* a function.

Functions can be used as mathematical models.

25. Allen Wilkins shoots a free throw in basketball. Sketch a reasonable graph showing its distance above the gym floor as a function of time since he shot the ball.

Making a model involves finding the particular equation for a function.

26. Find the particular equation of the linear function whose graph goes through the point $(-3, 5)$, and is parallel to the graph of $y = 7x - 13$. Write the answer in slope-intercept form.

27. Find the particular equation of the quadratic function containing the three points $(1, 16)$, $(2, 15)$, and $(4, 1)$.

Using the model involves finding x or y when the other is given.

28. If $f(x) = 3x + 7$, find $f(-4)$.

29. If $g(x) = 5x^2 - 3x + 6$, find $g(7)$.

30. If $y = 3x^2 - 5x - 2$, find x when $y = 11$.

Systems of linear inequalities and equations can be solved and graphed.

31. Plot the graph of the system:

$$y < -3x + 7$$
$$2x - 3y \le 12$$
$$y \ge -8$$

(Test 13 continued)

32. Write N_z, the *numerator* determinant for z, and evaluate it.

$$\begin{aligned} 3x + 5y + 2z &= -8 \\ 4x - y + 3z &= 7 \\ x + 2y + 5z &= 1 \end{aligned}$$

33. Solve by augmented matrices.

$$\begin{aligned} 4x + 7y &= 1 \\ 3x - 5y &= 11 \end{aligned}$$

Now it is time for you to show that you can put everything together to work a mathematical model problem.

34. The number of dollars Wolf Photo charges you to develop a roll of film and make prints varies linearly with the number of good pictures that are printed. For a 12-print roll they charge \$7.33, and for a 15-print roll they charge \$8.29.

a. Write the particular equation expressing dollars in terms of number of prints.
b. How much would a 36-print roll cost?
c. If your bill is \$12.77, how many prints did they make?
d. What does the dollars-intercept represent in the real world?
e. What do they they charge *per print*? How do you tell?
f. Sketch the graph.

Systems of linear inequalities can be used to find the optimum point in a linear programming problem.

35. The graph shows the feasible region for the manufacture of two kinds of tennis shoe, Plain (x pairs), and Deluxe (y pairs). The profit on each pair of Plains is \$20 and the profit on each pair of Deluxes is \$30.

a. Write an equation for the total profit in terms of x and y.
b. Shade the portion of the feasible region in which the profits is at least \$15,000.
c. How many pairs of each should be produced to make the maximum profit?
d. What is the maximum feasible profit?

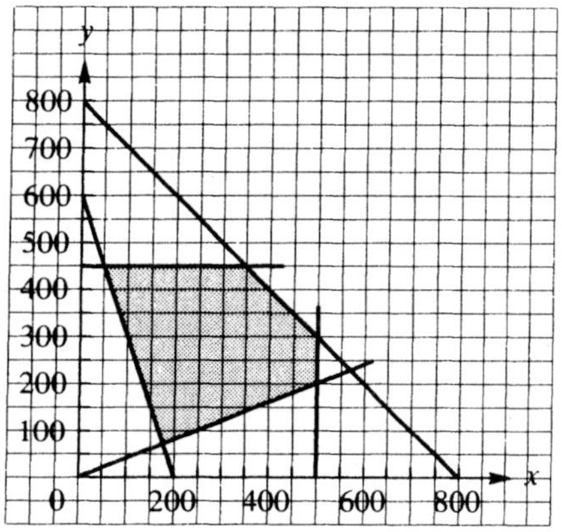

NAME: ______________________________ DATE: ____________

Test 14 | CHAPTER 6, Sections 6-1 to 6-5

Algebra/Trig is a study of functions. Recently you have been studying exponential functions.

1. Write the general equation that defines an exponential function.

2. Sketch the graph of the exponential function $f(x) = 2^x$ for values of x from -3 to 3.

3. Given the exponential function $f(x) = 3 \cdot 5^x$, evaluate $f(3)$, $f(-2)$, and $f(0)$.

To deal effectively with exponential functions you must know the properties of exponents. Complete the following.

4. $(ab)^k =$ **5.** $\dfrac{x^a}{x^b} =$ **6.** $(y^u)^p =$

7. The words "exponent," "base," and "power" apply to the symbol x^y. Tell which word goes with which part of the symbol.

Calculators can be used to raise numbers to non-integer powers. Evaluate the following.

8. $2^{3.7}$ **9.** $0.8^{-\frac{3}{7}}$

10. Explain why your answer to Problem 8 is *reasonable*.

Three definitions of exponentiation allow exponents to be zero, negative, or fractions. With these, you can operate on some powers *without* a calculator.

11. Use the definitions of exponentiation to show why $8^{-\frac{2}{3}} = 0.25$.

12. If $5^x = 625$, what does x equal?

13. Quick! What does 100^0 equal?

The properties of exponentiation let you transform expressions. Simplify the following.

14. $(4x^6y^{10})^3$ **15.** $\sqrt[6]{(11^{7.5})(11^{-5.1})}$

16. $(6x^{\frac{2}{3}})(5x^{-\frac{1}{5}}y^7)$ **17.** $\left(\dfrac{x^5}{y^6}\right)^7\left(\dfrac{y^8}{x^9}\right)^{10}$ **18.** $\dfrac{3 \cdot 2001^{501}}{5 \cdot 2001^{500}}$

It is important to be able to work old problems.

19. Write the general equation of a quadratic function.

20. Draw a number-line graph: $14 - 3x > 29$

NAME: ______________________________ DATE: ______________

Test 15 | CHAPTER 6, Sections 6-6 to 6-12

Algebra/Trig is a study of functions. Properties of exponential functions allow you to handle large and small numbers with scientific notation. Evaluate the following.

1. $(8.7 \times 10^{38})(9.324 \times 10^{-13})$ Round appropriately.

2. $\dfrac{3 \times 10^{871}}{5 \times 10^{265}}$ **3.** $(3 \times 10^{-7})^4$ **4.** $\dfrac{1}{2.5 \times 10^9}$

5. For the number 2.3×10^7, what special name is given to:
a. the numeral 2.3?
b. the numeral 7?

Logarithms allow you to find unknown exponents.

6. Solve for x: $4.86^{5x} = 3267$. Demonstrate that your answer is correct.

Logarithms have various properties.

7. Complete the following: $\log r^p =$

8. Demonstrate by calculator that $\log\left(\dfrac{57}{19}\right) = \log 57 - \log 19$.

9. Prove that $\log_b (rk) = \log_b r + \log_b k$.

10. Write as a single log of a single argument:
$\log_3 24 + \log_3 17 - \log_3 51$

11. Suppose that $\log_b 5 = 1.234$. Quick! What does $\log_b 25$ equal?

Demonstrate that you know the definition of logarithm by using it to find the *exact* value of x. No decimal approximations!

12. $\log_3 x = 5$ **13.** $\log_{32} 8 = x$ **14.** $\log_x 64 = \dfrac{2}{3}$

The definition of logarithm can be used to find logs to other bases.

15. Find $\log_7 451.9$.

Other functions besides exponentials have inverses.

16. Given the function $f(x) = 7x - 28$
a. Write an equation for $f^{-1}(x)$ in terms of x.
b. Evaluate $f^{-1}(35)$.
c. Evaluate $f(11)$.
d. Show that $f(f^{-1}(x)) = x$.

17. The sketch shows the graph of function g.
a. Draw the graph of g^{-1}.
b. Tell why g^{-1} is *not* a function.

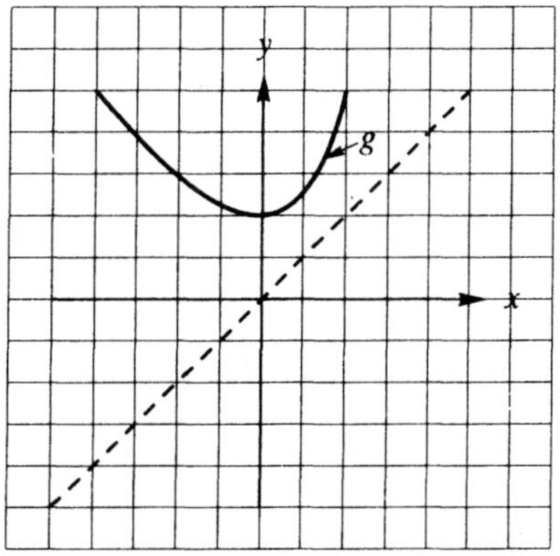

NAME: ______________________________ DATE: ______________

Test 16 | CHAPTER 6, Sections 6-13 and 6-14

Algebra/Trig is a study of functions. Recently you have used exponential functions as mathematical models. Doing so involves logarithms. Demonstrate that you know the three properties of logarithms by transforming the following. Do *not* evaluate!

1. $\log (3 \cdot 4) =$ **2.** $\log 3^4 =$ **3.** $\log \left(\frac{3}{4}\right) =$

Logarithms can be used to solve exponential equations.

4. Solve $3.7^{4x} = 9271$

The add-multiply property of exponential functions can sometimes be used to find values *without* finding the particular equation.

5. If f is an exponential function, and $f(5) = 20$ and $f(9) = 14$, use the add-multiply property to calculate $f(13)$, $f(17)$, and $f(21)$. Show your method.

6. Draw the graph of the function in Problem 5.

7. Show that the following data could *not* possibly be fit by an exponential function:

x	y
3	72
5	200
7	392

8. Show that y in Problem 7 varies directly with the square of x.

9. If $g(x)$ varies exponentially with x, and $g(0) = 37$ and $g(7) = 5940$, find the particular equation.

10. If $h(t) = 17.4 \times 1.23^t$, find $h(13)$.

11. If $z(t) = 60 \times 1.03^t$, find t when $z(t) = 144$.

An entire mathematical model application involves getting information from the words of the problem.

12. *Water Hyacinth Problem* The number of water hyacinth plants growing on a lake increases exponentially with time. On Tuesday the number of plants is 150. That Friday the number has risen to 240. If it takes 3400 plants to completely cover the lake, will be the lake be completely covered by 20 days after Tuesday? Justify your answer.

NAME: ______________________________ DATE: ______________

Test 17 | CHAPTER 6

Algebra/Trig is a study of functions. This chapter concerns exponential functions and their inverses, logarithmic functions.

1. Given $f(x) = x^3$ and $g(x) = 3^x$. One of the functions is an exponential function and the other is not. Which is which? How do you tell?

2. Evaluate: -7^4

3. Evaluate: $5 \cdot 8 + 2 \cdot 4^3$

4. Use variables to state the following:
a. Definition of logarithm
b. Property of a quotient of powers with equal bases.
c. Property of the log of a product
d. Definition of $x^{\frac{1}{n}}$.

5. Simplify: $(3r^5m^7)^4(64r^0m^{12})^{\frac{2}{3}}$

6. Evaluate $\sqrt[9]{32586.4}$

7. Evaluate in scientific notation. Round appropriately.

$$\frac{3.7495 \times 10^{-21}}{7.24 \times 10^{17}}$$

8. If $f(x) = 7^x$, write an equation expressing $f^{-1}(x)$ in terms of x.

9. Solve for x and simplify: $\log_{25} x = \frac{3}{2}$

10. Find $\log_6 37.9$.

11. Solve for x: $\log_x 64 = \frac{2}{3}$

12. Express as a single log of a single argument:

$$3 \log_5 12 - \log_5 36$$

13. Write the characteristic and mantissa of 8.75×10^{29}.

14. Suppose that $h(x)$ is an exponential function for which $h(2) = 24$ and $h(5) = 36$. Find $h(11)$.

15. Sketch the graph of $y = a \cdot b^x$ if $0 < b < 1$.

16. *Aspirin Problem* Assume that the concentration of aspirin in the blood system decreases exponentially with time after you take the tablets. At time $t = 0$ the concentration is 100 units, and at time $t = 40$ minutes the concentration is 73 units. Find the particular equation and use it to perdict the concentration at time $t = 95$ minutes.

NAME: ______________________________ DATE: ______________

Test 18 | CHAPTER 7, Sections 7-1 to 7-5

Algebra/Trig is a study of functions. In order to deal effectively with rational algebraic functions you must be good at factoring polynomials. Factor the following completely.

1. $12a^3b^5 + 18ac$

2. $9x^2 - 36$

3. $x^2 + 14xy - 15y^2$

4. $4x^2 + 2x - 30$

5. $4z^2 + 25z + 6$

6. $9x^2 - 48x + 64$

7. $x^3 + c^3$

8. $8p^3 - f^3$

9. $(x + a)y + (x + a)(3p - 4)$

Sometimes it is important to know whether or not a polynomial factors without actually finding the factors.

10. Prove that $36x^2 + 72x + 25$ is prime.

Factoring is the reverse process of multiplication. Multiply the following.

11. $(3x - 7)(4x + 8)$

12. $(x^2 - 7x + 3)(x - 4)$

13. $(5x + 11)^2$

14. $(x - 3)(x + 5)(x - 8)$

Long divide. Write the answer in mixed-number (or polynomial) form.

15. $\dfrac{4x^3 - 3x^2 + 7x - 19}{x - 2}$

16. $\dfrac{x^3 - 11}{x + 3}$

Sketch the graph, showing especially the behavior where the denominator equals zero.

17. $f(x) = \dfrac{1}{x - 3}$

18. $g(x) = \dfrac{(x + 2)(x - 3)}{x + 2}$

It is important to remember old concepts.

19. Sketch the graph of a quadratic function with a positive x^2-coefficient and a negative y-intercept.

NAME: ______________________________ DATE: ____________

Test 19 | CHAPTER 7, Sections 7-6 to 7-9

Algebra/Trig is a study of functions. In order to deal effectively with rational algebraic functions you must be good at operating with algebraic fractions. Perform the indicated operations and simplify.

1. $\dfrac{x^2 - 64}{x^2 - 16} \div \dfrac{x - 8}{x - 4}$

2. $\dfrac{x - 7}{7} - \dfrac{x - 4}{4}$

3. $\dfrac{x}{x + y} + \dfrac{y}{x - y}$

4. $\dfrac{5x + 17}{x^2 + 8x + 7} - \dfrac{3}{x + 7}$

5. $\dfrac{20 + \dfrac{10}{x}}{20 - \dfrac{5}{x^2}}$

To work effectively with rational expressions you must be able to factor polynomials.

6. Use the Factor Theorem to factor $x^3 - 5x^2 - 2x + 24$. Factor further, if possible.

7. Given $P(x) = 3x^3 + 13x^2 - 4x - 4$:

a. show that $P\left(\dfrac{2}{3}\right) = 0$.

b. Write a linear factor with integer coefficients. (It is not necessary to find the other factor.)

8. Factor: $x^7 - y^7$

9. Factor: $a^3 + b^3$

The purpose for operating with algebraic fractions is to be able to plot graphs of rational functions. Plot the following, showing in each case what happens at each place the denominator equals zero.

10. $f(x) = \dfrac{5x - 10}{x^2 + x - 6}$

11. $f(x) = \dfrac{x^2 - 3x - 4}{x + 1}$

It is important to keep up your skill with mathematical models.

12. Windy's sells large orders of french fries (5 3/4 ounces) for 89 cents and medium orders (4 ounces) for 75 cents. Assuming that the number of cents is a linear function of the number of ounces, write the particular equation. If a giant order cost $1.47, predict the number of ounces it would contain. Sketch the graph.

NAME: ______________________________ DATE: ______________

Test 20 | CHAPTER 7, Sections 7-10 and 7-11

Algebra/Trig is a study of functions. You have now studied variation functions, some of which are rational functions.

1. Write the general equation of a direct fifth-power variation function.

2. Sketch the graph of an inverse square variation function.

3. If p varies inversely with r, and $p = 37$ when $r = 54$, write the particular equation.

4. A variation function has $f(2) = 5$ and $f(6) = 45$. Which kind of variation function is it? Write the particular equation.

Variation functions can be used as mathematical models.

5. *Height-Weight Problem* The weights of people of the same proportions vary directly with the *cube* of the height. Phoebe Small is 5′ (i.e., 60″) tall and weighs 99 pounds.

a. Write the particular equation expressing weight as a function of height.

b. Predict the weight of a girl who is 6′4″ (i.e., 76″) tall, but who has the same proportions as Phoebe.

c. If Phoebe's little sister has the same proportions, but weighs only 80 pounds, how tall is she?

In working with rational fuctions you must sometimes solve fractional equations. Solve, discarding any extraneous solutions.

6. $1 + \dfrac{x}{x-3} = \dfrac{5}{x-3}$

7. $\dfrac{5}{x+1} + \dfrac{x}{x-2} = \dfrac{8x-7}{x^2-x-2}$

The purpose for operating with algebraic fractions is to facilitate plotting of rational function graphs.

8. Plot the graph, showing all critical features.

$$f(x) = \frac{12(x-1)(x-2)}{(x-1)(x+1)(x-4)}$$

NAME: ______________________________ DATE: ______________

Test 21 | CHAPTER 7

Algebra/Trig is a study of functions. Operating with rational functions involves simplifying algebraic fractions. Simplify these.

1. $\dfrac{(x+3)(x-5)}{(x-4)(x-1)} \div \dfrac{(x+3)(5-x)}{(x+4)(x+7)}$

2. $\dfrac{x}{x-2} - \dfrac{3}{x+4}$

3. $\dfrac{x^5 + y^5}{x + y}$

4. $\dfrac{x + \dfrac{2}{x+3}}{2 + \dfrac{x}{x+3}}$

5. Multiply: $(3x - 7)(x^2 - 4x + 5)$

Simplifying rational expressions involves factoring.

6. Factor completely: $x^3 + 125$

7. Factor completely: $64x^2 - 16$

8. Given: $P(x) = x^3 - 5x^2 - 8x + 48$

a. Show that $P(-3) = 0$.

b. Factor $P(x)$ completely.

An "improper" algebraic fraction can be transformed by long division.

9. Write in mixed-number form:

$$\frac{3x^3 - 14x^2 - 29x + 5}{x - 6}$$

Operation with rational functions sometimes involves solving fractional equations.

10. Solve: $x + \dfrac{1}{x-1} = \dfrac{x}{x-1}$

Variation functions can be used as mathematical models.

11. Write the general equation for

a. y varies inversely with the square of x.

b. y is directly proportional to the cube of x.

12. Find the particular equation if y varies directly with x, and $y = 63$ when $x = 9$.

It is important to remember other kinds of functions.

13. Find the particular equation of the exponential function containing (0, 234) and (6, 5810).

A purpose for operating with algebraic fractions is to facilitate plotting of rational function graphs.

14. Plot the graph, showing all critical features.

$$f(x) = \frac{12(x-1)}{(x-1)(x+2)}$$

NAME: ______________________________ DATE: ____________

Test 22 | CHAPTER 8

Algebra/Trig is a study of functions. Irrational algebraic functions involve radicals or non-integer exponents.

Write in simple *radical* form:

1. $\sqrt{18} + 2\sqrt{50} - \sqrt{98}$

2. Simplify: $\dfrac{21}{\sqrt{12} + \sqrt{5}}$

3. Simplify: $\dfrac{1 - \dfrac{1}{\sqrt{3}}}{1 + \dfrac{1}{\sqrt{3}}}$

Solving radical equations sometimes leads to extraneous solutions. Solve and check.

4. $\sqrt{x + 3} + \sqrt{x - 2} = 5$

Variation functions can involve non-integer exponents.

5. *Labor Problem* Assume that the length of time it takes to do a major construction project varies inversely with the 0.7 power of the number of workers on the project. Suppose that 100 workers can construct an office building in 250 days. Write the particular equation expressing time in terms of number of workers, and use it to predict the number of workers needed to construct the building in just 180 days.

Variation functions can involve more than two variables.

6. *Bridge Problem* Building a 200-foot-long bridge across Scorpion Gulch, where the water is 8 feet deep, costs $700,000. Assume that the cost varies directly with the length of the bridge and directly with the square of the depth of the water. Write the particular equation expressing cost in terms of length and depth. Use the equation to predict the cost of a bridge downstream from the present one, where the water is only 5 feet deep, but the length must be 600 feet.

Graphs of irrational functions show things about extraneous solutions. Let $f(x) = x - \sqrt{x - 2}$.

7. Find $f(11)$, $f(6)$, $f(3)$, $f(2.5)$, $f(2)$, and $f(0)$.

8. Plot the graph of f.

9. What is the domain of x if $f(x)$ is to be a *real* number?

10. Show that there are *no* real values of x for which $f(x) = 0$.

11. Set $f(x) = 4$, and solve for x. Show on your graph that one solution is valid, but the other one is extraneous.

NAME: ______________________________ DATE: ____________

Test 23 | CHAPTERS 6 through 8

(2 hours)

Algebra/Trig is a study of functions. This test measures your knowledge of exponential, rational, and irrational functions, and refreshes your memory about linears and quadratics.

1. Sketch the graph of
a. a decreasing exponential function.
b. an inverse variation function.
c. a rational algebraic function with vertical asymptote at $x = -2$ and removable discontinuity at $x = 4$.
d. an increasing linear function.
e. a quadratic function with negative x^2-coefficient.

2. Write the general equation for a function that has the indicated property.
a. Quadrupling x makes y 64 times as big.
b. Halving x makes y 4 times as big.
c. Adding a constant to x multiplies y by a constant.
d. Adding a constant to x adds a constant to y.

3. Multiply and simplify.
a. $(\sqrt{8} + \sqrt{3})(\sqrt{15} + \sqrt{2})$
b. $(\sqrt{3} - \sqrt{5})^2$
c. $(\sqrt{y} - 6)(\sqrt{y} + 6)$
d. $(3x + 11)(2x - 7)$
e. $(x + 3)(x^2 - 4x - 5)$

4. Factor completely.
a. $a^5 + b^5$
b. $5x^2 + 27x + 36$
c. $x^3 + 2x^2 - 11x - 12$
d. $64p^2 - 36$

5. Carry out the indicated operations and simplify.
a. $\dfrac{1}{x^2 + 4x + 3} + \dfrac{1}{x^2 - 1}$

b. $\dfrac{9 - x^2}{3x + 9} \div \dfrac{x^2 - 5x + 6}{x^2 + 2x + 1}$

6. Simplify the following. *No* decimal approximations!
a. $5x^{-\frac{2}{5}} \cdot 7x^{\frac{3}{4}}$

b. $\sqrt{125} \div \sqrt[3]{25}$

c. $\dfrac{12}{\sqrt{19} + 4}$

d. $\dfrac{24}{\sqrt[3]{2}}$

NAME: ______________________________ DATE: ______________

(Test 23 continued)

e. $\sqrt[6]{36}$

f. $7\sqrt{45} + \dfrac{10}{\sqrt{5}}$

g. $\log_9 27$

h. $\log_7 30 + 2 \log_7 5$

i. $\left(\dfrac{5}{x - 2} + 1\right) \div \left(\dfrac{5}{x + 3} - 1\right)$

7. Solve the equation.

a. $3^{2x} = 457$

b. $2 + \sqrt{x - 2} = x$

c. $x + \dfrac{1}{x - 2} = \dfrac{x - 3}{2 - x}$

8. Given $f(x) = \dfrac{x^3 + 2x^2 - 11x - 12}{x^2 - x - 6}$, tell the ordered pair at which there is a removable discontinuity.

9. *Bacteria Problem* The number of bacteria in a bottle of milk increases exponentially with time. Assume that there are 56 million bacteria at time $t = 0$ days, and 204 million at $t = 2$ days. Find the particular equation and use it to predict the number of bacteria at $t = 7$.

10. *Braking Distance Problem* The number of feet it takes your car to stop after you apply the brakes varies directly with the square of your speed and inversely with the coefficient of friction between your tires and the roadway. Assume that on a dry road the coefficient of friction is 0.8, and it takes 45 feet to stop when the speed is 30 miles per hour.

a. Write the particular equation expressing braking distance in terms of speed and coefficient of friction.

b. Lisa Carr left 290 feet of skid marks in coming to a stop on a wet road where the coefficient of friction was only 0.3. Predict her speed when she applied the brakes.

11. *Christmas Tree Problem* Data on the price of Christmas trees from Maple Grove Fruit Farm in Montague, Michigan, show that the price is directly proportional to the 2.3 power of the height of the tree. A 6-foot Douglas fir costs $48.00. Write the particular equation expressing price in terms of height. Use the equation to predict the price of a 60-foot tree such as might be used at the White House.

NAME: ____________________ DATE: ____________

Test 24 | CHAPTER 9, Sections 9-1 to 9-5

Algebra/Trig is a study of functions and relations. Quadratic relations have graphs that are conic sections.

Transform the equation or inequality by completing the square. Sketch the graph.

1. $x^2 + y^2 + 4x - 10y + 20 < 0$

2. $4x^2 + y^2 - 8x + 6y - 3 = 0$
(Also, show the foci.)

3. $y^2 + 2x - 6y - 7 = 0$

You can reverse the graphing process and find the equation for a given conic.

4. Write an equation for the hyperbola sketched. Transform to the form $Ax^2 + Cy^2 + Dx + Ey + F = 0$, where A, C, D, E, and F are integer constants.

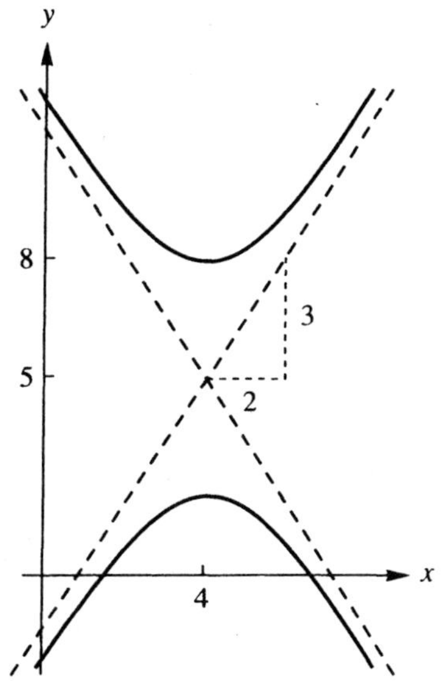

It is important to remember old techniques.

5. Solve: $\dfrac{-9}{x - 2} - x = 8$

6. Solve: $\sqrt{x^2 + 35} - x = -7$

7. Simplify: $\dfrac{22}{\sqrt{18} + \sqrt{7}}$

NAME: ______________________________ DATE: ______________

Test 25 | CHAPTER 9, Sections 9-6 to 9-8

Algebra/Trig is a study of functions and relations. Equations of quadratic relations can be found from geometrical information.

1. For each point in a set of points, its distance from $(-2, 4)$ is three times its distance from the line $x = 6$.
   a. Find the equation.
   b. Tell which conic section the graph will be.

If the equation of a quadratic relation has an xy-term, you can use the discriminant to tell which conic it will be.

Find the discriminant. Tell which conic section each graph will be.

2. $4x^2 + 9y^2 + 7x - 2y - 100 = 0$

3. $4x^2 + 10xy + 9y^2 + 7x - 2y - 100 = 0$

4. $4x^2 + 12xy + 9y^2 + 7x - 2y - 100 = 0$

5. $4x^2 + 14xy + 9y^2 + 7x - 2y - 100 = 0$

6. $4x^2 + 14xy + 7x - 2y - 100 = 0$

Intersections of two conics can be found by solving the system of equations.

7. Solve the system

$$7x^2 - 5y^2 + 20y = 3$$
$$21x^2 + 5y^2 = 209$$

8. Solve the system and sketch the graphs.

$$x^2 + y^2 - 4x + 2y = 20$$
$$4x + 3y = 5$$

NAME: ______________________ DATE: __________

Test 26 | CHAPTER 9

Algebra/Trig is a study of functions and relations. In this chapter you have learned that quadratic relation graphs are conic sections.

1. The graph of $x^2 + y^2 = 36$ is a circle. How can you tell this without actually plotting?

2. Where is the center of the circle in Problem 1?

3. For each of the following, make *one* change in the equation from Problem 1 so that its graph will

a. be the region outside the circle.
b. have a larger radius.
c. have its center somewhere else.
d. be an ellipse.
e. be a hyperbola.
f. be rotated so that its axes are not parallel to the coordinate axes.

4. Find the points where these conics intersect.

$$16x^2 - 5y^2 = 64$$
$$16x^2 + 25y^2 - 96x = 256$$

5. You have learned how to sketch the conic from a given equation. You can use the same knowledge to get the equation from the graph. Write the particular equation of this ellipse. (It is not necessary to transform it.)

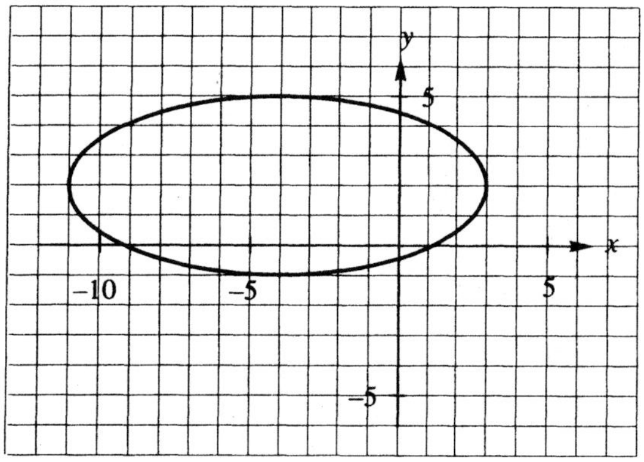

6. For each point in a set of points, its distance from (3, 4) is four times its distance from (−5, 2).

a. Find the equation.
b. Tell which conic section the graph will be.

7. Sketch the graph: $x = \frac{1}{4}y^2 - 2y - 1$

8. Sketch the graph: $x^2 - 9y^2 = 36$

NAME: ______________________________ DATE: ______________

Test 27 | CHAPTER 10, Sections 10-1 to 10-3

Algebra/Trig is a study of functions and relations. Quadratic and higher degree functions sometimes involve complex numbers.

Perform the indicated operations and write in simple radical form.

1. i^{59} **2.** i^{-15} **3.** i^{500}

4. $\sqrt{-18} \times \sqrt{-25}$ **5.** $\sqrt{(-5)(-35)}$

6. $\dfrac{\sqrt{98}}{\sqrt{-2}}$ **7.** $\sqrt{-12} + \dfrac{12}{\sqrt{-3}}$

Perform the indicated operations and simplify.

8. $(5 - 2i) + (7 + 6i)$ **9.** $(-8 + 2i) - (3 - 11i)$

10. $(11 + 4i)(10 - 3i)$ **11.** $(12 + i)(12 - i)$

12. $(12 + i)^2$ **13.** $\dfrac{12 + 9i}{3 - 4i}$

14. Factor over the set of complex numbers: $x^2 - 6x + 58$

Solve. Simplify the solutions.

15. $x^2 + 10x + 29 = 0$ **16.** $3x^2 - 6x + 4 = 0$

17. $16x^2 + 49 = 0$

18. Check one of the solutions for Problem 15.

Write a quadratic equation with real coefficients having the following solutions. The final answer should be in the form $ax^2 + bx + c = 0$.

19. Solutions are 3 and -5.

20. One solution is $6 + 4i$.

It is important to remember old techniques.

21. Tell which conic section the graph will be:

$$x^2 - y^2 + 3x - 5y + 13 = 0$$

NAME: ______________________________ DATE: ______________

Test 28 | CHAPTER 10, Sections 10-4 to 10-6

Algebra/Trig is a study of functions and relations. Higher degree functions have graphs with varying numbers of vertices and intercepts.

1. Sketch the graph of
   a. a seventh degree function.
   b. a quadratic function with no real zeros.
   c. a quadratic function with exactly one real zero.

Synthetic substitution can be used to evaluate polynomial functions and to help determine their properties.

2. If $P(x) = 3x^5 - 2x^4 - 5x^3 - 17x^2 + 4x - 11$, find $P(2)$ by synthetic substitution.

3. According to the Upper Bound Theorem, could $P(x)$ in Problem 2 have any zeros that are greater than 2? Justify your answer.

4. According to Descartes' Rule of Signs, how many positive, negative, and non-real complex solutions could $P(x)$ in Problem 2 have?

5. Although you may never have done this, you can use synthetic substitution for *complex* values of x. Do this to evaluate $P(1 + i)$ if

$$P(x) = x^3 + x^2 - 4x + 7.$$

It is important to remember old techniques.

6. Find the particular equation of the linear function that contains (3, 10) and (7, −2). Transform to $Ax + By = C$ form, where A, B, and C are integer constants.

Higher degree functions can be used as mathematical models.

7. Find the particular equation of the cubic function containing (2, 16), (3, 30), (4, 36), and (5, 28).

8. *Airplane Payload Problem* The "payload" an airplane will carry is a *cubic* function of the length of the plane. Suppose that for a particular design of plane the equation is

$$P(L) = -L^3 + 5L^2 - 3L - 1,$$

where $P(L)$ is the number of tons of payload, and L is the length of the plane in *tens* of meters.
   a. Evaluate $P(L)$ for each integer value of L from −1 through 5 and use the results to plot the graph.
   b. $P(L)$ has an *integer* zero. What is it?
   c. Find all other zeros of P.
   d. An airplane will fly only if the payload it can carry is positive. Darken the portion of your graph which represents planes that will fly.
   e. Approximately how long is the longest plane that will fly? Justify your answer.

NAME: ______________________________ DATE: ______________

Test 29 | CHAPTER 10

Algebra/Trig is a study of functions and relations. In this chapter you have studied higher degree functions and complex numbers.

Perform the indicated operations and simplify.

1. i^{37}

2. i^{-31}

3. $\sqrt{-12} \times \sqrt{-15}$

4. $(3 + 9i) - (12 - 5i)$

5. $\dfrac{8 + 6i}{7 - 5i}$

Complex numbers can appear as solutions of quadratic equations, or in factors of polynomials.

6. Solve: $x^2 - 6x + 13 = 0$

7. Factor over {complex numbers}: $9x^2 + 49$

Special properties relate the sum and product of quadratic equation solutions to the coefficients.

8. For $3x^2 - 7x + 2 = 0$
  a. Solve the equation.
  b. Verify that the sum of the solutions is $-\dfrac{b}{a}$.
  c. Verify that the product of the solutions is $\dfrac{c}{a}$.

9. Write a quadratic equation with real coefficients if one of the solutions is $-5 + 7i$.

Using higher degree functions as mathematical models involves finding the particular equation from given ordered pairs.

10. Write the particular equation of a quartic function that could, according to Descartes' Rule of Signs, have four positive zeros.

11. Find the particular equation of the cubic function that contains $(-2, -35)$, $(1, 7)$, $(2, 9)$, and $(3, 25)$.

Higher degree function graphs can be sketched or plotted.

12. Sketch the graph of a quintic function with exactly three distinct real zeros.

13. Given: $P(x) = x^3 - 3x^2 - 5x + 10$
  a. Evaluate $P(x)$ for each integer value of x from -3 through 5.
  b. Use the results to plot the graph of P.
  c. Find all zeros of $P(x)$.
  d. Prove that 5 is an upper bound on the zeros of $P(x)$.

NAME: ______________________________ DATE: ____________

Test 30 | CHAPTER 11, Sections 11-1 to 11-6

Algebra/Trig is a study of functions. Sequences are functions whose domains are the set of positive numbers, and series are indicated sums of the terms of sequences.

1. For the sequence 3, 8, 15, 24, 35, 48, 63, ...
   a. write the next two terms.
   b. write an equation for t_n in terms of n.

2. a. Find t_{73} for the arithmetic sequence 4, 7, 10, ...
   b. Find S_{73} for the arithmetic series $4 + 7 + 10 +, \ldots$.
   c. Find n if t_n equals 1075 in the arithmetic sequence 4, 7, 10, ...
   d. Find n if $S_n = 23{,}001$ for the arithmetic series $4 + 7 + 10 + \ldots$.

3. a. Find a decimal approximation for t_{15} in the geometric sequence with first term 83 and common ratio 1.07.
   b. Find a decimal approximation for S_{15} in the geometric sequence with first term 83 and common ratio 1.07.
   c. Find n if t_n is approximately 10,832 in the geometric sequence with first term 83 and common ratio 1.07.
   d. Find n if S_n is approximately 15,408.15.

4. a. Write three arithmetic means between 4 and 324.
   b. Write three geometric means between 4 and 324.

5. Evaluate: $\sum_{k=2}^{7} (-1)^k(k - 1)$

Geometric series with common ratios between -1 and 1 *converge* as the number of terms gets very large.

6. Find the number to which this geometric series converges:
$200 - 170 + 144.5 - \ldots$

7. Write this repeating decimal as a ratio of relatively prime integers:
0.324324324...

It is important to remember about old topics such as imaginary numbers.

8. Write the first 8 terms of the geometric sequence with first term 10 and common ratio $2i$.

Since sequences are functions, they can be graphed.

9. Sketch the graph of the geometric sequence with first term 24 and common ratio 1/2.

NAME: ______________________________ DATE: ______________

Test 31 | CHAPTER 11, Section 11-7

Algebra/Trig is a study of functions. You should be comfortable operating with arithmetic and geometric sequences and series, either with the formulas or without, whichever is appropriate.

1. A sequence has $t_1 = 5$ and $t_2 = 15$.
   a. If the sequence is geometric, what is the common ratio? What do t_3 and t_4 equal?
   b. If the sequence is arithmetic, what is the common difference? What do t_3 and t_4 equal?

2. An arithmetic series has first term 8 and third term 20. What does the fifth term equal? What does the third partial sum equal?

3. A geometric series with first term 7 converges to 70. What does the common ratio equal? What does the second partial sum equal?

4. An arithmetic series begins $3 + 8 + \ldots$. Is 23 a term in this series? If so, which term? If not, why not?

Sequences and series are good mathematical models if the independent variable takes on only *discrete* values rather than varying continuously.

5. *Compound Interest Problem* If you invest \$500 in a savings account that pays 8% per year interest, compounded quarterly, how much will you have in the account at the end of 11.6 years?

6. *Oil Field Problem* Assume that the amounts of oil produced by the first, second, third, ... , wells in a new oil field form a decreasing geometric sequence. If the first well produces 80,000 barrels and the second produces only 70,000 barrels, predict the total amount of oil that could be produced from the field if many wells are drilled.

7. *Swing Problem* When a person is swinging, the amplitude (the swing's distance above its rest position) increases with each pump. According to *Scientific American* magazine (April, 1977, page 60) the amplitude increases by *adding* a constant when the person pumps sitting down, and by *multiplying* by a constant when the person pumps standing up. Suppose that when you get on a swing, its amplitude is 3 cm (i.e., $t_0 = 3$). Assume that the amplitude increases by 2 cm each pump when you are sitting, and increases by multiplying by 1.2 each pump when you are standing.
   a. What is your amplitude after 20 pumps if you are standing? If you are sitting?
   b. What is the minimum number of pumps needed to make the amplitude equal 75 cm when you pump sitting down? Standing up?
   c. If the amplitude is 0 when you first get on the swing, explain why you would have to start pumping sitting down, not standing up.

NAME: ______________________________ DATE: ____________

Test 32 | CHAPTER 11, Sections 11-8 to 11-10

Algebra/Trig is a study of functions. The value of a term in a binomial series is a function of the term number.

1. Expand as binomial series and simplify.

a. $(r - p)^9$ b. $(2x + y^3)^5$ c. $(1 + i)^6 \quad (i = \sqrt{-1})$

2. Find the terms. Leave coefficients as factorials.

a. Term with t^5 in $(v - t)^{17}$
b. Term with x^{10} in $(x + y)^{19}$
c. Seventh term of $(p - j)^{55}$
d. Term with b^8 in $(a + b^2)^{10}$

3. The numbers 1 7 21 35 35 21 7 1 form a row of Pascal's Triangle.

a. Which power of $(a + b)$ does the row go with?
b. Write the next row.

4. Write the following in compact form using factorials.

a. $38 \cdot 37 \cdot 36 \cdot 35 \cdot 34 \cdot 33$
b. $\dfrac{14!}{19 \cdot 18 \cdot 17 \cdot 16 \cdot 15}$

5. Simplify the following expressions.

a. $\dfrac{8!}{4!}$ b. $\dfrac{11!}{3!8!}$ c. $\dfrac{(n-1)!}{n!}$

6. The quantity 0! is defined to equal 1. Tell why this definition is *reasonable*.

7. For what values of n will $n!$ be exactly divisible by 7?

8. In the binomial series from $(a + b)^5$, the powers of a decrease by 1 each term, and the powers of b increase by 1. If you carry on this pattern beyond the "end" of the series you get

$$a^5 + \_\,a^4b + \_\,a^3b^2 + \_\,a^2b^3 + \_\,a^1b^4 + \_\,a^0b^5 + \_\,a^{-1}b^6 + \_\,a^{-2}b^7 + \ldots$$

where the spaces are for the coefficients.

a. Write the coefficients for the seventh and eighth terms, and tell how you you get them.
b. What will be the coefficients for terms beyond 8? How do you know?

NAME: ______________________ DATE: ____________

Test 33 | CHAPTER 11

Algebra/Trig is a study of functions. In this chapter you have studied sequences and series.

1. For the arithmetic series 13 + 19 + ... ,
   a. find t_{27}.
   b. find S_{27}.
   c. if $t_n = 445$, find n.

2. For the geometric series 50 + 48 + ... ,
   a. 867.373 is the approximate value of a partial sum of the series. How many terms were added to get this number?
   b. To what number does the series converge?

3. Write four arithmetic means between 35 and 98.

4. a. Evaluate $\sum_{k=1}^{5} 2^k + 3$.
   b. Is this series arithmetic, geometric, or neither? Justify your answer.

5. For the sequence 0, 3, 8, 15, 24, 35, ... ,
   a. write the next two terms.
   b. find a formula for t_n in terms of n.

6. Expand as a binomial series: $(p^2 + 3h)^4$

7. Write the term of the binomial series. Simplify.
   a. The term with b^5 in $(a - b)^9$
   b. The seventh term of $(x^{\frac{1}{2}} + x^{-\frac{1}{2}})^8$.

8. *Office Building Problem* Suppose that you are responsible for predicting the cost of constructing a new office building. You find that the first story costs \$500 per square meter of floor space, and the second story costs \$600 per square meter. Assume that the costs per square meter form a geometric sequence.
   a. Find the cost per square meter for the 10th story.
   b. If each story has an area of 1000 square meters, find the total cost of the first 10 stories of the building.
   c. Find the maximum height of the building if the top story is to cost no more than \$9000 per square meter.
   d. Sketch the graph of cost per square meter for the first 10 stories.

It is important to remember old techniques and concepts.

9. Sketch the graph of a cubic function that has one real zero and two non-real complex zeros.

NAME: ______________________________ DATE: ______________

Test 34 | CHAPTER 12, Sections 12-1 to 12-5

Algebra/Trig is a study of functions. The probability of an event is a function of the number of outcomes in that event. Counting numbers of outcomes can involve permutations and combinations.

1. How do you tell whether a given counting problem involves *permutations* or *combinations*?

2. What is the difference between an *outcome* of a random experiment and an *event* of that experiment?

Compute.

3. The number of combinations of 8 elements taken 3 at a time.

4. The number of permutations of 8 elements taken 3 at a time.

5. ${}_9C_6$ **6.** ${}_{11}P_4$ **7.** ${}_5C_5$

8. Using the letters in ADJUSTING without repeats,

a. How many different four-letter arrangements could be made?
b. How many different nine-letter arrangements could be made?
c. How many of these nine-letter arrangements have a consonant in fifth position and J in sixth position?
d. What special name is given to the sixth position in part (c)?
e. If a nine-letter arrangement is selected at random, what is the probability that a consonant is fifth and J is sixth?

9. Twenty people apply for seven jobs that are available at Fly-By-Night Aircraft Manufacturing Co.

a. In how many different ways could such a group of seven people be composed?
b. Twelve of the applicants are men and the rest are women. How many of these seven-person groups have exactly three men?
c. How many of these seven-person groups have exactly three men *or* exactly three women?
d. What is the probability that a group of seven selected at random will have exactly three men or exactly three women?
e. After the seven people have been selected, in how many different ways could five of them be assigned to the Assembly Department?
f. In how many different ways could five of the seven people be assigned to the five available positions on the assembly line?

NAME: ______________________________ DATE: ____________

Test 35 | CHAPTER 12

Algebra/Trig is a study of functions. In this chapter you have learned that probability is the dependent variable in a function of a random variable.

1. What is meant by the sample space for a random experiment?

2. A card is drawn at random from a normal 52-card deck. Find the probability that it is either black or a king.

3. Find the number of different ways 9 people from a group of 13 could be arranged in a line.

4. Find the number of different groups that could be made using 11 people from a set of 30.

5. Sketch the normal distribution curve.

6. *Sawmill Problem* Records from Buzz Sawmill show that workers have a 0.02 probability of losing an arm and a 0.03 probability of losing a leg in any one year.

a. What is the probability that a worker

i. does *not* lose an arm? ii. does *not* lose a leg?

iii. loses neither an arm nor a leg? iv. loses both an arm and a leg?

b. Armand Legg Casualty Insurance Company plans to offer insurance policies for the workers. The company will pay $300 for the loss of an arm, $400 for the loss of a leg, and $7000 for the loss of both an arm and a leg. On the average, how much will the company expect to pay out per policy?

c. Armand Legg wants to make a profit of $5 per policy. How much should the premium be? That is, how much should the company charge each worker for the policy?

7. *Claire's ESP Problem* Claire Voyance takes a five-question true-false test. She hasn't studied, so she guesses at random. By using her ESP she figures she has a 60% probability of getting any one answer right.

a. What is the probability of getting an answer wrong?

b. Let $P(x)$ be the probability of getting exactly x of the five answers right. Calculate $P(0)$ through $P(5)$.

c. Show that your answers to part (b) are reasonable by finding $P(0$ or 1 or 2 or 3 or 4 or $5)$.

d. What is the probability of getting at least three answers right?

e. What special name is given to this kind of probability distribution, and why is it given that name?

f. Plot the graph of this probability distribution.

8. *Average Grade Problem* Miss Calculate gives a test to the 10 students in her first-period algebra class. The grades are

93, 82, 95, 86, 90, 84, 89, 79, 87, 93

She wishes to predict how well her later classes will do on the same test.

a. Find the mean and standard deviation of the 10 scores.

b. List all the reasons you can think of for why the later classes might have a different mean or standard deviation.

NAME: ______________________________ DATE: ____________

Test 36 | CHAPTERS 9 through 12

(2 hours)

Algebra/Trig is a study of functions and relations. In these chapters you have studied functions and relations involving quadratics, higher degree polynomials, sequences and series, and probability (random variables).

1. Write an inequality for the elliptical region sketched. Transform it to a polynomial inequality with integer coefficients.

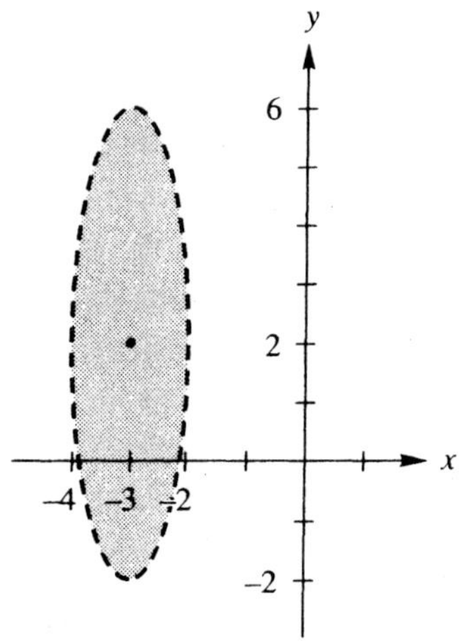

2. Sketch the graph of

$$4x^2 - 9y^2 - 16x - 18y = 29$$

3. Solve the system:

$$16x^2 + 3y^2 - 32x + 6y - 72 = 0$$
$$2x + y = 8$$

4. A set of points has the property that each point is equidistant from the line $x = 3$ and the point $(5, 2)$. Find an equation. Transform it to a polynomial equation with integer coefficients. Which conic section will the graph be?

5. For the polynomial function graphed, tell the degree, the number of real zeros, and the number of non-real complex zeros.

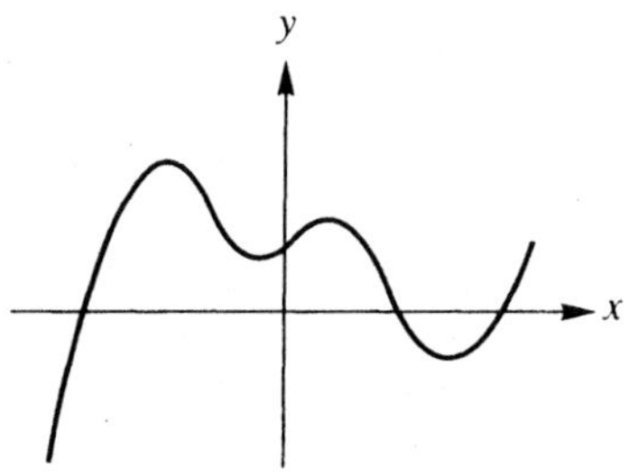

6. If $f(x) = x^3 - 7x^2 + 17x - 15$

a. Find $f(4)$ by synthetic substitution.

b. Show that 3 is a zero of $f(x)$.

c. Find all other zeros of $f(x)$.

d. Show that the number of positive zeros agrees with the conclusion of Descartes' rule of signs.

e. Show how the Upper Bound Theorem insures that there are no zeros of $f(x)$ which are greater than 8.

7. Simplify: i^{-23}

8. Simplify: $\dfrac{50}{3 + 4i}$

9. Write a quadratic equation with real coefficients for which $2 + 3i$ is one solution.

NAME: ______________________________ DATE: ____________

(Test 36 continued)

10. Find the particular equation of the cubic function that contains (3, 42), (2, 14), (1, 4), and (0, 0).

11. Write a formula for the nth term, t_n, of a sequence if the first terms of the sequence are 1, 3, 7, 15, 31, 63, 127, 255.

12. Find t_{71} for the arithmetic sequence 5, 9, 13, ...

13. Find S_{17} for the geometric series with first term 81 and common ratio 1.09.

14. Write three arithmetic means between 6 and 486.

15. *Nail Problem* Assume that when a nail is pounded into a board, the number of centimeter it moves on each pound form a geometric sequence.

a. If a particular nail moves 0.8 cm on the first pound and 0.7 cm on the second pound, find the *total* distance the nail will have moved after many pounds.

b. Sketch the graph of the sequence.

16. Find the term that contains b^{10} in the binomial series that comes from expanding $(a - b^2)^8$.

17. One card is drawn at random from a normal deck of 52 cards. What is the probability that it is a black face card?

18. Calculate ${}_{12}P_5$.

19. Using the letters QRSTUVWXYZ, how many different arrangements could be made if X, Y, or Z must come first?

20. How many different committees could be composed using 3 boys and 4 girls if there are 10 boys and 9 girls from whom to choose?

21. *Sports Team Problem* The varsity tennis team has an 80% probability of winning the district championship. The soccer team has only 70% probability of winning its district championship. What is the probability that at least one of the teams wins its district championship?

22. *Grades Problem* Juan DeMony has a probability of 0.6 for making an A in each of three courses. His parents will reward him as follows:

3 A's: $50
2 A's: $10
1 A: $3
0 A's: He pays them $20.

What is Juan's mathematically expected payoff?

23. Find the mean and standard deviation of: 57, 43, 49, 51, 55.

NAME: ______________________________ DATE: ____________

Test 37 | CHAPTER 13, Sections 13-1 to 13-4

Algebra/Trig is a study of functions. Trigonometry and circular functions have angles or arcs of circles as their independent variables.

1. As you ride a Ferris wheel, your distance from the ground is a function of time. Sketch a reasonable graph. Tell the special name given to this kind of function.

Find the reference angle. Draw the angle in standard position, and show the reference angle.

2. $193°$ **3.** $-290°$ **4.** $146° 13'$ **5.** $2841°$

Sketch the indicated arc of a unit circle in standard position.

6. $\frac{2\pi}{3}$ **7.** $\frac{\pi}{5}$ **8.** π **9.** 2

For Problems 10 through 16, find the *exact* value.

10. The six trigonometric functions of $-120°$.

11. $\sec 180°$ **12.** $\cot 6360°$ **13.** $\sin^2 60°$

14. $\cos \pi$ **15.** $\tan \frac{3\pi}{4}$ **16.** $\csc 0$

For Problems 17 through 22, find a 4-significant-digit decimal approximation.

17. $\sin 217°$ **18.** $\tan 24° 37'$ **19.** $\sec 82°$

20. $\cos 0.345$ **21.** $\cot 1$ **22.** $\csc 2$

23. Find the radian measure of an angle of $150°$.

It is important to be able to use concepts backwards as well as frontwards.

24. Find θ if $\cos \theta = -\frac{\sqrt{3}}{2}$ and θ is between $90°$ and $180°$.

25. Find $\sin \theta$, if θ is in standard position and its terminal side contains the point $(7, -4)$.

While working with new kinds of functions, you must remember old ones.

Write the general equation of

26. a linear function. **27.** a quadratic function.

NAME: ______________________ DATE: ____________

Test 38 | CHAPTER 13, Sections 13-5 to 13-8

Algebra/Trig is a study of functions. Trigonometric and circular functions have graphs that are periodic.

1. Sketch the graph: $y = 10 + 6 \sin 5(\theta - 10°)$

2. Sketch the graph: $y = \tan \theta$

3. Sketch the graph: $y = \sec x$

4. Write an equation for this sinusoid.

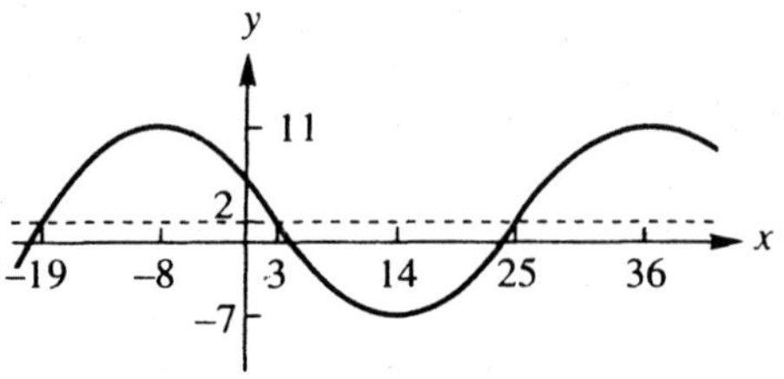

5. The distance between Earth and Mars is (approximately) a sinusoidal function of time whose equation is

$$d = 228 + 93 \cos \frac{\pi}{390}(t - 560)$$

where t is the number of days after today, and d is in millions of miles.

a. Sketch the graph.
b. How far apart are the two planets today?
c. What is the farthest they are apart? The closest?
d. How much time elapses between times of closest approach?
e. (Bonus!) How long does it take Mars to make a complete orbit around the sun?

6. Write the *exact* number of radians for 270°.

7. Write the *exact* number of degrees for $\frac{2\pi}{3}$ radians.

Find the *exact* value.

8. $\cos \frac{\pi}{4}$ **9.** $\sin \frac{4\pi}{3}$ **10.** $\cot \frac{\pi}{2}$

11. tan 120° **12.** csc 135° **13.** sec 180°

14. $\sec \theta$, if $(-2, 3)$ is on the terminal side.

15. Find and sketch the reference angle for 3504°.

NAME: ______________________ DATE: ____________

Test 39 | CHAPTER 13, Sections 13-9 to 13-11

Algebra/Trig is a study of functions and relations. The inverses of the circular functions are relations.

Find the exact value(s).

1. $\text{Cos}^{-1} \frac{\sqrt{3}}{2}$ **2.** $\arcsin \frac{1}{2}$ **3.** $\text{Arctan}\,(-1)$

4. $\text{Arccot}\,(-1)$ **5.** $\text{Sec}^{-1} 2$ **6.** $\text{Sin}^{-1} 2$

7. $\tan\left(\text{Cos}^{-1} \frac{12}{13}\right)$ **8.** $\csc\,(\text{Arccsc}\,\sqrt{31})$

9. Given: $y = 6 + 8 \cos \frac{\pi}{7}(x - 4)$

a. Find y when $x = 17$.
b. Transform the equation so that x is in terms of y.
c. Find the first three positive values of x for which $y = 11$.

10. As you ride a bicycle, the distance between your foot and the pavement varies sinusoidally with the horizontal distance the bicycle has gone. Suppose that you start with your right foot somewhere between a high point and a low point, and push down. When you have gone 5 meters, your right foot first reaches its lowest point, 11 cm above the pavement. The high points are 45 cm above the pavement. The bicycle moves a horizontal distance of 16.4 meters for each complete revolution of the pedals.

a. Sketch the graph of the sinusoid.
b. Write a particular equation of the sinusoid.
c. What was the second time your foot was 43 cm above the pavement?

11. Sketch the graph of $y = \cos^{-1} x$. Darken the part corresponding to $y = \text{Cos}^{-1} x$.

It is important to apply your knowledge to problems that you have not seen before.

12. On the same set of axes, using the same scales on each axis, draw graphs of $y = \tan x$ and $y = \text{Arctan}\, x$. Remember that $\pi \approx 3.14$. Then draw the line $y = x$. What relationship do you notice among the three graphs?

NAME: ______________________________ DATE: ____________

Test 40 | CHAPTER 13

Algebra/Trig is a study of functions and relations. In this chapter you have studied trigonometric and circular functions, and their inverses.

Find the *exact* values.

1. $\cos 45°$ **2.** $\cot \frac{2\pi}{3}$ **3.** $\text{Sec}^{-1}(-\sqrt{2})$

Find the decimal approximations.

2. $\sin 63° 19'$ **5.** $\csc 1.2$ **3.** Arctan 3

7. Find an angle between 0° and 90°, inclusive, whose cotangent is undefined.

8. Find a real number between 0 and $\frac{\pi}{2}$, inclusive, whose cosine is $\frac{\sqrt{3}}{2}$.

9. Sketch the graph of $y = \tan x$.

10. Sketch the graph of $y = \sin^{-1} x$.

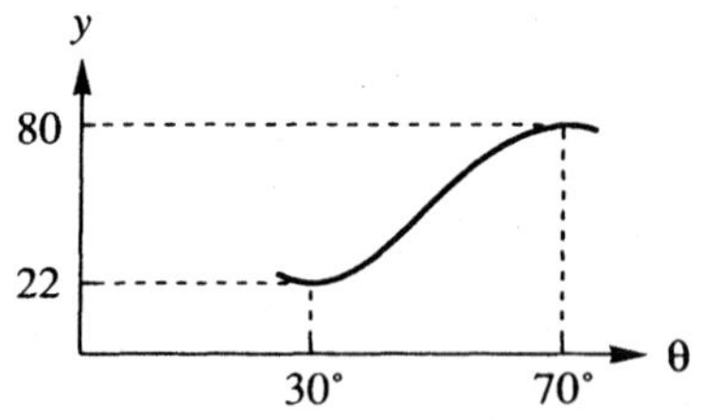

11. Write the particular equation of this sinusoid.

12. *Breathing Problem* Assume that as you breathe, the volume of air in your lungs is a sinusoidal function of time. Suppose that at time $t = 2$ seconds your lungs have their maximum volume, 3.2 liters. When you exhale, your lungs still have 0.6 liters. Your breathing rate is one complete breath every 5 seconds.

a. Sketch the graph of volume versus time.
b. Write an equation expressing volume in terms of time.
c. How much air did you have in your lungs at time $t = 0$?
d. Between what two positive values of t did your lungs first have no more than 1 liter of air?

NAME: ______________________ DATE: ______________

Test 41 | CHAPTER 14, Sections 14-1 and 14-2

Algebra/Trig is a study of functions and relations. You are learning properties that will help you solve trigonometric equations.

Prove that each of the following equations is an identity.

1. $(1 - \tan A)^2 = \sec^2 A - 2 \tan A$

2. $\cos^3 B \sin^2 B = \cos^3 B - \cos^5 B$

3. $\dfrac{1}{1 - \cos C} + \dfrac{1}{1 + \cos C} = 2 \csc^2 C$

4. $\csc D \cos^2 D + \sin D = \csc D$

5. $(5 \cos E - 2 \sin E)^2 + (2 \cos E + 5 \sin E)^2 = 29$

6. $\dfrac{\csc F}{\cos F} - \dfrac{\cos F}{\sin F} = \tan F$

NAME: ______________________________ DATE: ______________

Test 42 | CHAPTER 14, Sections 14-3 and 14-4

Algebra/Trig is a study of functions and relations. You are learning properties that will help you solve trigonometric equations where the argument is more complicated than x.

1. Write the odd-even property for sine.

2. Write the odd-even property for cosine.

3. Write the cofunction property for tangent.

4. If $\cos x = -\frac{2}{3}$ and x terminates in Quadrant II, find the exact value of $\cos 2x$, $\sin 2x$, and $\tan 2x$.

5. Write an equation expressing $\tan 12x$ in terms of $\tan 6x$ alone.

6. Derive a triple argument property expressing $\cos 3x$ in terms of $\cos x$ alone.

7. Find the exact value of $\tan 15°$ by using the composite argument property for $\tan (A + B)$, and clever choices for A and B. Write the answer in simple radical form.

8. Prove that the following is an identity:

$$\sin(\theta + 60°) + \cos(\theta + 30°) = \sqrt{3} \cos \theta$$

9. Solve the equation for values for x greater than or equal to 0 and less than 2π. Write the general solution first.

$$\cos^2 x - \sin^2 x = \frac{\sqrt{3}}{2}$$

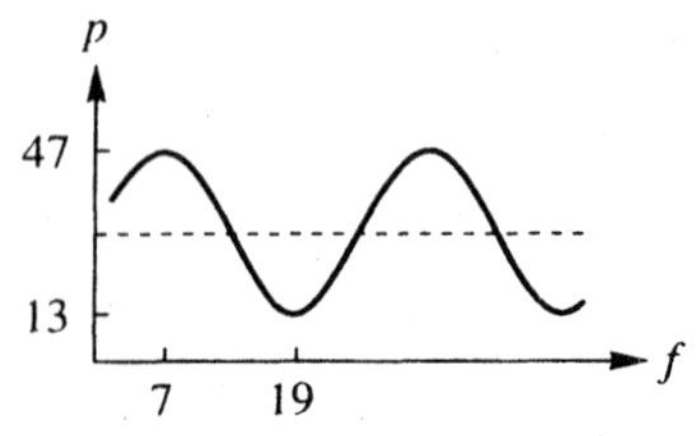

10. Write a particular equation for the sinusoid sketched.

NAME: ______________________________ DATE: ______________

Test 43 | CHAPTER 14, Sections 14-5 to 14-8

Algebra/Trig is a study of functions and relations. You are learning properties that will help you solve trigonometric equations.

If A is between 270° and 360°, and $\cos A = \frac{12}{13}$, find the *exact* value of:

1. $\sin 2A$ **2.** $\cos 2A$ **3.** $\sin \frac{1}{2}A$ **4.** $\cos \frac{1}{2}A$

Write equations expressing

5. $\cos 2x$ in terms of $\cos x$ and $\sin x$.

6. $\cos 2x$ in terms of $\cos x$ alone.

7. $\cos 2x$ in terms of $\sin x$ alone.

8. $\tan \frac{1}{2}x$ in terms of $\cos x$ alone.

9. $\tan \frac{1}{2}x$ in terms of $\cos x$ and $\sin x$.

10. $\cos 37° \cos 99°$ as a sum.

11. $\sin 4 + \sin 7$ as a product.

12. $\cos 8x - \cos 2x$ as a product.

13. $\cos 5x \sin 9x$ as a sum.

14. $19 \cos(\theta - 50°)$ as a sum of sine and cosine.

15. $8 \cos \theta + 15 \sin \theta$ as a single sinusoid.

Solve the equation in the given domain.

16. $\frac{1 - \cos \theta}{\sin \theta} = \sqrt{3}, \quad 0° \le \theta < 360°$

17. $\cos 4x + \cos 2x = 0, \quad 0 \le x < \pi$

NAME: ____________________ DATE: ____________

Test 44 | CHAPTER 14

Algebra/Trig is a study of functions and relations. In this chapter you have learned properties of the trigonometric and circular functions that help you solve equations.

1. Express $\cos^2 x$ in terms of $\sin x$.

2. Express $\cos 2x$ in terms of $\sin x$.

3. Write a product of two trig functions that equals 1.

4. Write the odd-even property for an odd trig function.

5. Write the cofunction property for $\cot \theta$.

6. Write $\cos 3x + \cos 7x$ as a product of functions with positive arguments.

7. If $\cos \theta = -\frac{2}{3}$ and $450° < \theta < 540°$, find

a. the exact value of $\cos \left(\frac{1}{2}\right)\theta$.

b. the exact value of $\tan 2\theta$.

8. Find the exact value of $\sin 15°$.

9. Prove that $\frac{\sec G}{\sin G} - \frac{\sin G}{\cos G} = \cot G$ is an identity.

10. Solve: $\sin 2x \sin x - \cos 2x \cos x = 1, x \in [-\pi, \pi]$

11. Solve: $\sqrt{2} \sin D \cos D = \sin D, D \in [-180°, 180°]$

12. Solve: $\tan (2\theta - 34°) = \sqrt{3}, \theta \in$ {real numbers of degrees}

13. The graph of $y = 7 \cos \theta + 24 \sin \theta$ would be a sinusoid if it were plotted. What would be the first positive value of θ at which there is a high point?

NAME: ____________________ DATE: __________

Test 45 | CHAPTER 15, Sections 15-1 to 15-6

Algebra/Trig is a study of functions. Trigonometric functions can be used to predict unknown lengths, angles, and areas of triangles.

Note: Capital letters are used both for an angle and for its measure.

1. Triangle ABC has $B = 137°$, $a = 31.6$ in., and $c = 42.8$ in. Find b.

2. Triangle DEF has $D = 58°$, $E = 41°52'$, and $f = 0.032$ microns. Find d.

3. Triangle GHI has $i = 19$ m, $H = 137°$, and $I = 51°$. Find g.

4. Triangle JKM has $j = 28$ cm, $k = 17$ cm, and $m = 13$ cm.
  a. Find the measure of the smallest angle.
  b. Find the area of the triangle.
  c. Let s be the "semi-perimeter" of triangle JKM. That is, s is $\frac{1}{2}$ the perimeter. Evaluate the expression

  $\sqrt{s(s - j)(s - k)(s - m)}$.

  What interesting observation do you notice?

5. Triangle NPQ has $q = 59$ miles, $P = 76°$, and $Q = 103°$. Find n.

6. RST is a right triangle with right angle S. If $s = 21.3$ Ångstroms and $t = 8.71$ Ångstroms, find the measure of the larger acute angle.

7. Triangle UVW is to be constructed with $U = 23°$, $v = 10$ cm, and $w = 7$ cm.
  a. Predict the length of u.
  b. Construct the triangle using angle U and sides v and w. Then measure u and show that it is within ± 0.1 cm of the predicted value.

8. Write the double-argument property for sine.

NAME: ______________________________ DATE: ______________

Test 46 | CHAPTER 15, Sections 15-7 and 15-8

Algebra/Trig is a study of functions. Trigonometric functions can be used to analyze vector quantities.

1. If $|\vec{r}| = 13$ and $|\vec{s}| = 19$, and the angle between $\vec{r}$ and $\vec{s}$ is 146°, find $|\vec{r} + \vec{s}|$, $|\vec{r} - \vec{s}|$, and the angle that each of these resultant vectors makes with $\vec{r}$.

2. A displacement vector is formed by adding a vector 2.3 cm long due west (270°) to another vector 4.8 cm long on a bearing of 130°.

a. Calculate the resultant magnitude and bearing.

b. Confirm your answer by accurate graphing.

3. Express $v = 29$ at $\theta = 214°$ as a sum of components in the x- and y-directions.

4. Express $4(3\vec{i} - 5\vec{j}) + 7(-\vec{i} + 2\vec{j})$ as a magnitude and direction.

5. Solve the equation: $\cos \theta + \sin \theta = 1$, $\theta \in [-270°, 90°]$

6. Find the exact value of $\cos\left(\text{Tan}^{-1}\frac{7}{24}\right)$.

NAME: ______________________ DATE: __________

Test 47 | CHAPTER 15

Algebra/Trig is a study of functions. The properties you have learned in this chapter allow you to use trigonometric functions to solve real-world problems involving triangles.

1. Rock star Nellie Wilson has a coat specially tailored with gold-embroidered lapels 37 cm by 33 cm by 13 cm (see sketch).

a. Find the measure of the largest angle of this triangular lapel, correct to the nearest minute.

b. Find the area of the lapel.

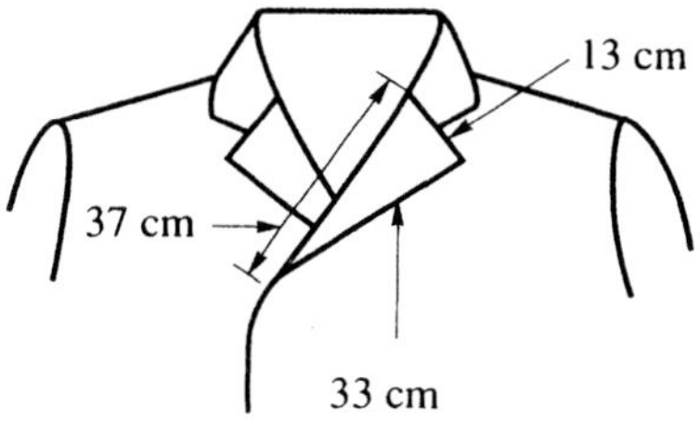

2. A ski lift is to be built to the top of Bald Mountain. From the base of the lift, the angle of elevation of the top of the mountain is 37°. From a point 185 ft further away from the mountain, the angle of elevation is only 29°43′. Find the length of the ski lift.

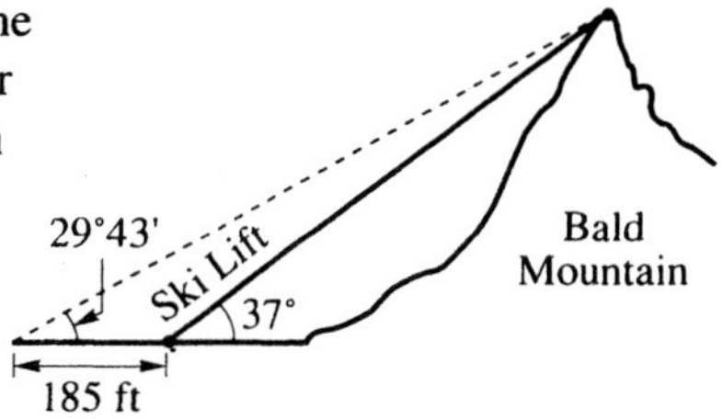

3. An airplane is flying due north at a speed of 200 miles per hour (mph). The wind is blowing 60 mph toward the north east, at an angle of 42° east of north. The plane's actual velocity with respect to the ground is the vector sum of the plane's velocity through the air and the wind's velocity. Find this resultant vector.

4. Highway 1984 and Farm Road 2001 intersect at a 32° angle. A CB radio operator is on Highway 1984, 5 km from the intersection (see sketch). The radio can pick up signals from as far away as 4 km. Between what two distances, d_1 and d_2, on Farm Road 2001 could signals be picked up by the CB radio on Highway 1984?

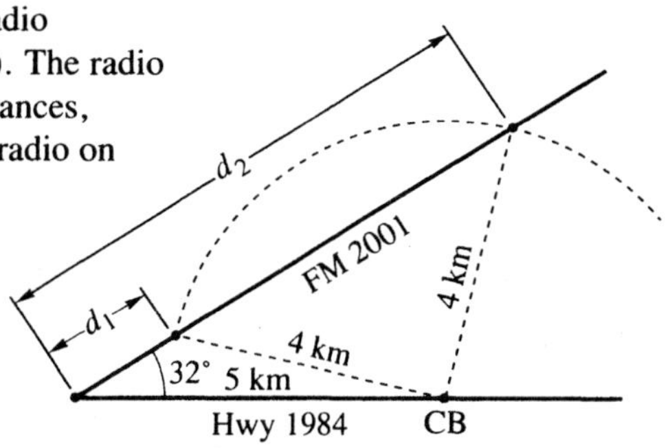

5. Find the period of the sinusoid $f(x) = 7 + 19 \cos \frac{\pi}{13}(x - 11)$.

NAME: ____________________ DATE: ____________

Test 48 | CHAPTERS 13 to 15

(2 hours)

Trigonometry is a study of a particular class of functions, the trigonometric functions, and another closely-related class of functions, the circular functions. In studying any kind of function you do four things: (1) define it; (2) graph it; (3) learn its properties; and (4) use it as a mathematical model. This test is to see how well you can do these four things.

1. Draw a sketch which clearly shows that you know the difference in definition between the argument of a trig function and the argument of a circular function.

2. Write the definition of $\sin\theta$ and the definition of $\sin x$.

3. For circular functions, the other four functions are defined in terms of $\sin x$ and $\cos x$. Write the definition of $\tan x$.

4. The inverse circular functions are defined in terms of the circular functions. Write the definition of $x = \cos^{-1} t$.

5. What is the difference between the definitions of $\operatorname{arcsec} x$ and $\operatorname{Arcsec} x$?

6. The definitions can be used to find *exact* values of the functions and inverses. Find the exact values of:

a. $\csc 240^\circ$ b. $\tan \frac{5\pi}{6}$ c. $\sec \pi$ d. $\cot 0$

e $\sin\theta$, if $\cot\theta = -\frac{8}{15}$, and θ terminates in Q. II.

f. $\operatorname{Sec}^{-1}\left(-\frac{2}{\sqrt{3}}\right)$, inverse *circular* function.

g. $\operatorname{Cos}^{-1} 0$, inverse *circular* function.

7. Sketch the graph:
a. $y = \tan\theta$ b. $y = \csc x$ **c.** $y = \cos^{-1} x$ d. $y = \operatorname{Cot}^{-1} x$

8. Values of the inverse trig functions are defined based on graphical considerations. The values of $\cot(-45^\circ)$ and $\cot 135^\circ$ are both equal to -1. Although -45° is closer to zero, $\operatorname{Cot}^{-1}(-1)$ is defined to be 135°. Explain why.

9. The graph of $f(x) = \sin x \cos x$ can be sketched by first using the double-argument properties. Sketch the graph.

10. Once you know the effects of the constants in a sinusoidal equation you can go backwards from the graph and find the particular equation. Find a particular equation for the circular function sketched.

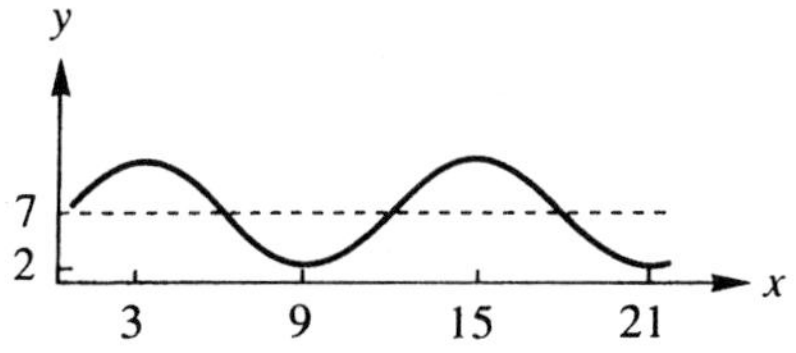

NAME: ______________________________ DATE: ____________

(Test 48 continued)

11. The techniques of Problem 10 allow you to use sinusoidal functions as mathematical models. Suppose that y varies sinusoidally with x according to the equation

$$f(x) = 8 + 11 \cos \frac{\pi}{7}(x - 2)$$

a. Find $f(19)$.

b. Find the first three positive x-intercepts.

12. The trig and circular functions have more properties than any functions you have studied so far. Write

a. The Pythagorean property for tangent.
b. The reciprocal property for secant.
c. The odd-even property for cosecant.
d. The cofunction property for the circular cosine.
e. The composite-argument property for $\cos(A - B)$.
f. The double-argument property for sine.
g. $\sin 5x \cos 7x$ as a sum, with positive arguments.
h. $\cos 37° + \cos 95°$ as a product of functions of positive angles.
i. $5 \sin \theta + 12 \cos \theta$ as a single cosine with a phase displacement.

13. The half-argument property for sine is tricky to use. Find the *exact* value of $\sin \left(\frac{1}{2}\right)\theta$ if $\cos \theta = \frac{4}{5}$ and $360° < \theta < 450°$.

14. One way to show that you know the properties is to prove identities. Prove that the following is an identity.

$$\sin x(\csc x - \sin x) = \cos^2 x$$

15. One reason for learning the properties is to solve equations. Solve $\cos^2 \theta - \sin^2 \theta = \frac{\sqrt{3}}{2}$ for θ in the domain $(0°, 360°)$.

16. One application of trig functions is analysis of triangles. Write the law of sines.

17. Triangle ABC is to be constructed with $a = 6.5$ cm, $b = 10$ cm, and $c = 4.5$ cm.

a. What will be the measure of the largest angle?
b. Confirm your prediction by accurate construction.

18. Sketch the *two* possible triangles XYZ if $x = 12$, $y = 10$, and $m\angle Y = 25°$. It is not necessary for you to calculate any sides or angles.

19. Trig functions enable you to make mathematical models of vector quantities. If $\vec{a} = 3\vec{i} + 2\vec{j}$, and $b = -7\vec{i} + 4\vec{j}$, find $\vec{a} + \vec{b}$ as

a. a sum of two components.
b. a magnitude and angle in standard position.

NAME: ______________________________ DATE: ______________

Test 49 | FINAL EXAMINATION

(3 hours)

Algebra/Trig is a study of functions and relations. This examination measures how well you know the definitions, graphs, properties, and applications of various elementary functions.

1. a. Write an equation illustrating the associative axiom of multiplication.
b. Write an example of a quadratic trinomial.
c. For the inequality $|3x - 14| < 19$, solve and draw the number-line graph if the domain is {integers}.

2. a. *Traffic Problem* The number of cars on the Golden Gate Bridge is a function of the time of day. Sketch a reasonable graph of the function.
b. Sketch the graph of a relation that has domain 1 to 7, inclusive, and range 3 to 5, inclusive, but which is *not* a function.

3. *Bookbinding Problem* The number of dollars per book charged for printing and binding is a linear function of the number of pages in the book. One manufacturer charges \$6.60 for a 160-page book and \$11.08 for a 384-page book. Write the particular equation for this function, and use it to predict the charge for printing and binding an 800-page book.

4. a. *Tax Bracket Problem* Federal income tax is charged at different rates for families with different incomes, x. In 1987 the tax was $f(x) = 330 + 0.15(x - 3000)$ for families with moderate incomes, and $g(x) = 4080 + 0.28(x - 28{,}000)$ for people with higher incomes. The tax rate changes where $f(x) = g(x)$. Find the value of the income, x, at which the change takes place.
b. Plot the graph of the system of inequalities.
$2x - y > 0$
$x + y \geq 3$
$y + 1 > \frac{1}{3}(x - 3)$

5. a. Find the particular equation of the quadratic function with vertex at $(3, -4)$, and x-intercept 1. Express the final answer in the form $y = ax^2 + bx + c$. Sketch the graph.
b. Solve the equation $x^2 - 8x + 41 = 0$ over the set of complex numbers.

6. An exponential function has $f(0) = 200$ and $f(5) = 100$. Find $f(15)$. Find the particular equation. Sketch the graph. Find x if $f(x) = 3$.

7. Plot the graph of the rational function, showing all critical features.

$$f(x) = \frac{6x^2 + 12x - 90}{x^3 + 4x^2 - 11x - 30}$$

8. a. Write in simple radical form: $\dfrac{92}{\sqrt{3} + 7}$
b. *Animal Body Problem* The skin area of a particular kind of animal is directly proportional to the 2/3 power of its mass. Assume that an 80-gram kitten has a skin area of 150 cm^2. Write the particular equation expressing area in terms of mass. Use the equation to find out what mass of cat has 2000 cm^2 of skin.

NAME: ______________________________ DATE: __________

(Test 49 continued)

9. a. Sketch the graph: $25x^2 + 4y^2 + 150x - 32y + 189 < 0$
b. Solve the system:

$$x^2 + y^2 - 4x = 21$$
$$x^2 + 5y^2 = 49$$

10. a. Sketch the graph of a sixth degree function that has two distinct real zeros and four non-real complex zeros.
b. Find all zeros of $f(x) = x^3 - 12x^2 + 46x - 52$.
c. Show that the number of positive zeros is in agreement with the conclusion of Descartes' Rule of Signs.

11. a. Find t_{78} for the arithmetic sequence 3894, 3826, ...
b. Find S_{60} for the geometric series with first term 300 and common ratio 1.03.
c. Find the number to which the geometric series $100 + 90 + 81 + \ldots$ converges.
d. Find the term with b^5 in the binomial series from expanding $(a - b)^{11}$.

12. a. Find the number of permutations of 7 elements taken 3 at a time.
b. Find the probability of getting a face card in a single draw from a normal 52-card deck.
c. If five dice are rolled, find the probability that exactly 3 of them show the number 6.
d. Find the mean and standard deviation of 37, 35, 40, 38, and 40.

13. a. For the sinusoid sketched, find the particular equation and use it to find the first three positive values of x for which $y = 5.3$.
b. Sketch the graph of $y = \sec \theta$.

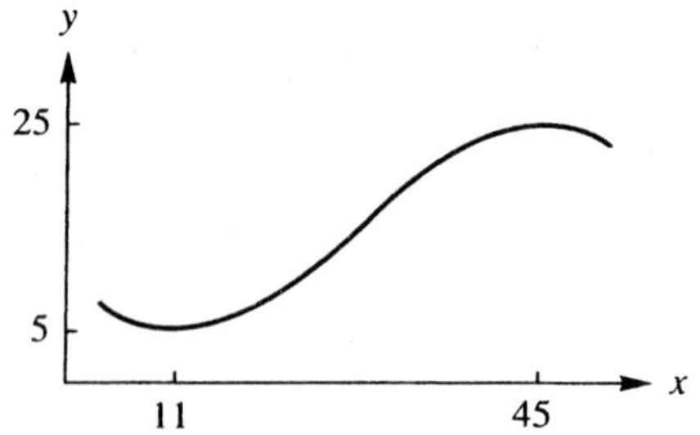

14. a. Write the Pythagorean property for secants and tangents.
b. Solve the equation $\cos^2 x - \sin^2 x = 0.5$ if x is between $-\pi$ and π.

15. a. A triangle is to be drawn with sides of 8, 9, and 16 units. Find the measure of the largest angle, and find the area.
b. A triangle is to be drawn with angles of 37° and 71°, and included side 7.8 cm. Find the length of the side opposite the 37° angle.

TEST 1 CHAPTER 1

1. example: $-\frac{1}{2}$

2. example: $\sqrt{2}$

3. example: $\sqrt{-1}$

4. $35.21 = \frac{3521}{100}$, which is a ratio of two integers.

5. a. $3x^2 - 6x + 7$
$= 3(5^2) - 6(5) + 7$
$= 75 - 30 + 7$
$= \underline{\underline{52}}$

b. $3x^2 - 6x + 7$
$= 3(-4)^2 - 6(-4) + 7$
$= 48 + 24 + 7$
$= \underline{\underline{79}}$

6. $3x - 7 = -40$
$3x = -33$
$x = -11$
a. $S = \{-11\}$
b. $S = \{-11\}$
c. $S = \varnothing$

7. $|2 - 3x| > 17$
$2 - 3x > 17$ or $2 - 3x < -17$
$-3x > 15$ or $-3x < -19$
$x < -5$ or $x > 6\frac{1}{3}$

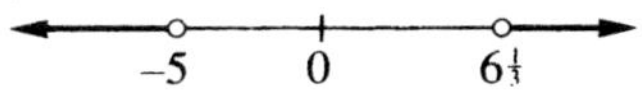

8. 9th degree binomial

9. Quartic trinomial

10. example: x^2

11. example: $\sqrt{x}$ or $\frac{1}{x}$

12. $20 - 12 \div 3 \cdot 2 + 1$
$= 20 - 8 + 1$
$= \underline{\underline{13}}$

13. $|14 - 31| - 8$
$= |-17| - 8$
$= \underline{\underline{9}}$

14. $7 - 5[4 - 3(2 + x)]$
$= 7 - 5[4 - 6 - 3x]$
$= 7 - 5[-2 - 3x]$
$= 7 + 10 + 15x$
$= \underline{\underline{17 + 15x}}$

15. $5\left[2x - \frac{1}{4}(20x - 3)\right]$
$= 5\left[2x - 5x + \frac{3}{4}\right]$
$= 5\left[-3x + \frac{3}{4}\right]$
$= \underline{\underline{-15x + 3\frac{3}{4}}}$

16. $(3x - 2)(5x + 4)$
$= \underline{\underline{15x^2 + 2x - 8}}$

17. $x^2 - 11x - 12$
$= \underline{\underline{(x - 12)(x + 1)}}$

(*Note*: Factoring is introduced informally via the "Do These Quickly" problems. It is presented formally in Chapter 7.)

18. If x and y are real numbers, then $xy = yx$.

19. If x, y, and z are real numbers, then $(x + y) + z = x + (y + z)$.

20. Every real number x has a real number $-x$ such that $x + (-x) = 0$.

21. If x and y are real numbers, then xy is a unique real number.

22. Commutativity for addition

23. Distributivity, multiplication over addition

24. If $x = y$, then $xz = yz$.
Proof:

$xz = xz$	Reflexive axiom
$x = y$	Given
$\therefore xz = yz$, Q.E.D.	Substitution into a product.

TEST 2 CHAPTER 2

1. $y = 0.2x^3$, $\{-3 \le x \le 3\}$

x	y
−3	−5.4
−2	−1.6
−1	−0.2
0	0
1	0.2
2	1.6
3	5.4

Range:
$\underline{\underline{\{y: -5.4 \le y \le 5.4\}}}$

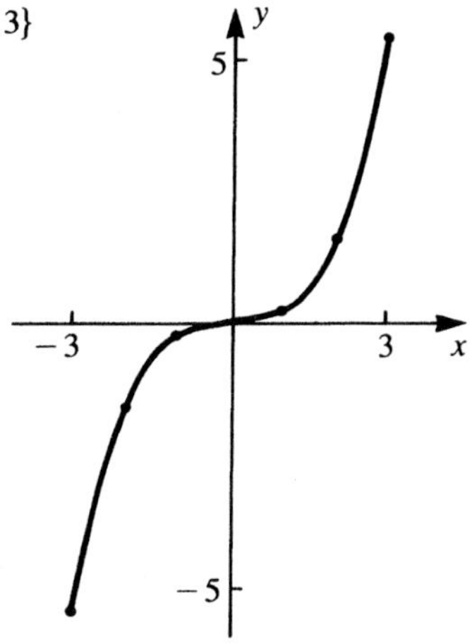

2. $y = x - 1$, $\{-2 < x < 4\}$

x	y
−2	−3
−1	−2
0	−1
1	0
2	1
3	2
4	3

Range: $\underline{\underline{\{y: -3 < y < 3\}}}$

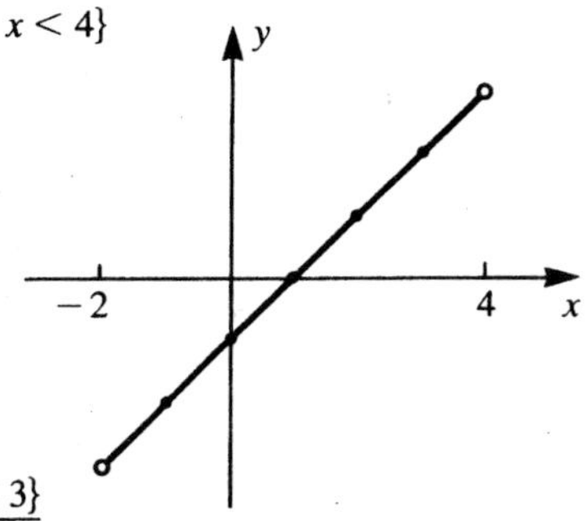

3. $y = |x - 3|$, {digits}

x	y
0	3
1	2
2	1
3	0
4	1
5	2
6	3
7	4
8	5
9	6

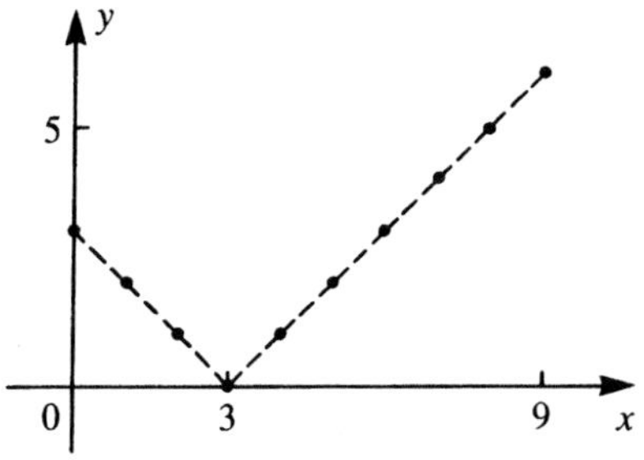

Range: $\underline{\underline{\{0, 1, 2, 3, 4, 5, 6\}}}$

4. Domain: $\{x: -5 \le x \le 7\}$
Range: $\{y: -4 \le y \le 6\}$
Function.

5. Domain: $\{x: -6 < x < 8\}$
Range: $\{y: -3 < y < 5\}$
Not a function.

6. $|y| = x + 2$
Domain: $\{x: x \ge -2\}$
Not a function.

x	y
-2	0
-1	± 1
0	± 2
1	± 3
2	± 4

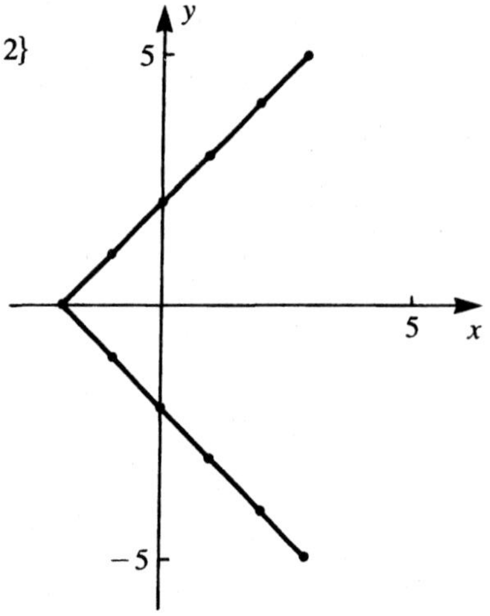

7. 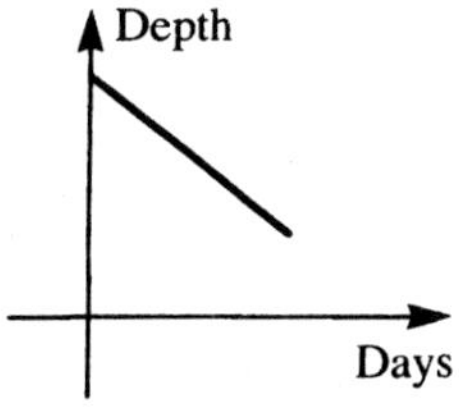

8. 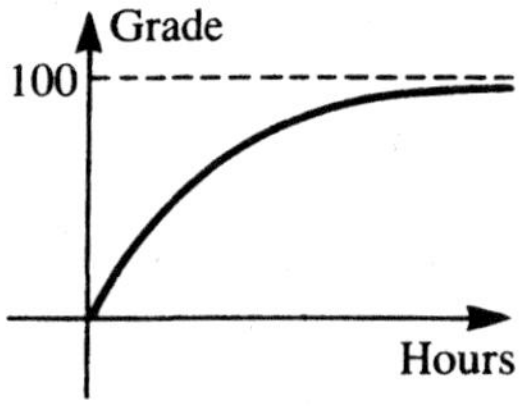

9. 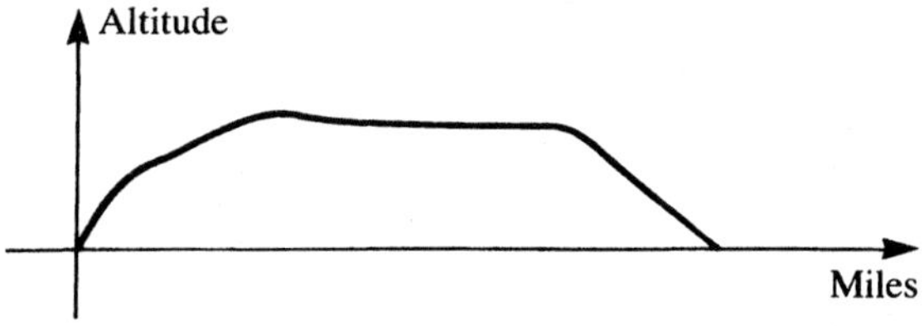

10. Example:

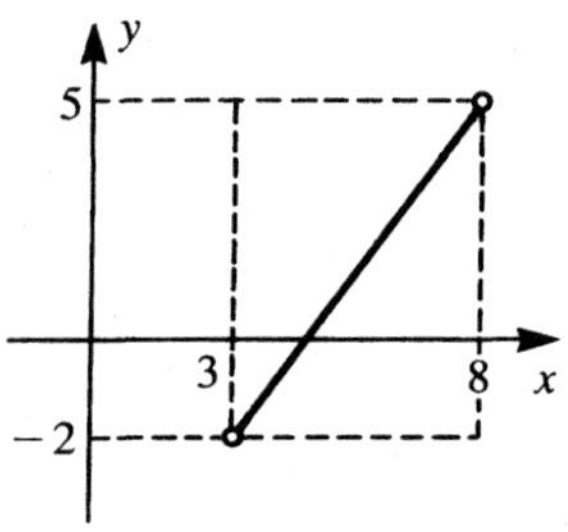

11. Example:

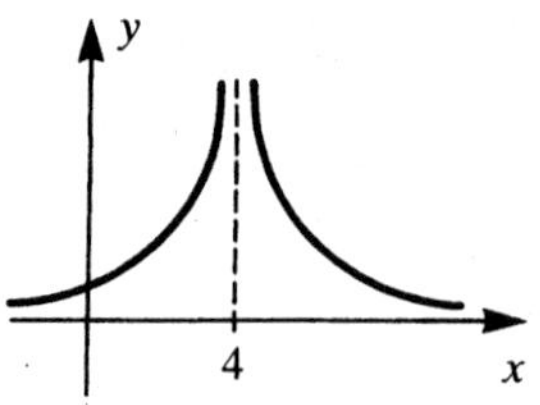

12. $|x + 5| \ge 2$
$x \ge -3$ or $x \le -7$

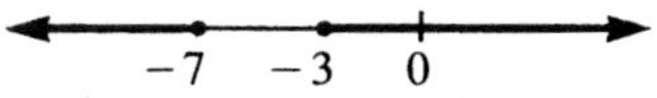

13. $7^3x^4y^5 + z^8$: 9th degree binomial

14. $\frac{3}{4} + \frac{1}{3} = \frac{9}{12} + \frac{4}{12} = \frac{13}{12}$ or $1\frac{1}{12}$

15. $7 - 5[4 - 3(2 + x)]$
$= 17 + 15x$

16. $(x + 3)(2x - 5) = 0$
$x + 3 = 0$ or $2x - 5 = 0$
$x = -3$ or $x = 2.5$
$S = \{-3, 2.5\}$

TEST 3 CHAPTER 3 Sections 3-1 to 3-4

1. $y = mx + b$

2. $3x - 7y = 42$
$y = \frac{3}{7}x - 6$

3. $y - 5 = -\frac{1}{2}(x - 3)$
$y = -\frac{1}{2}x + 6.5$

4. An x-intercept is a value of x when y is zero.

5. $3x - 5y = 90$
If $x = 0$, then $y = -18$.

6. $4x + 7y = 56$
If $y = 0$, then $x = 14$.

7. $y = -\frac{2}{5}x + 3$

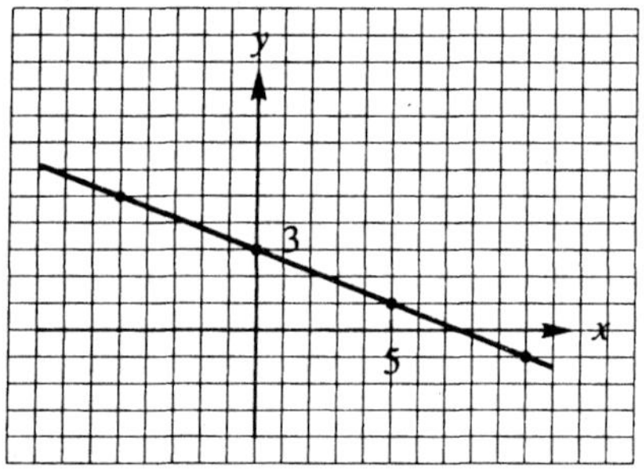

8. $y + 6 = 2(x - 4)$

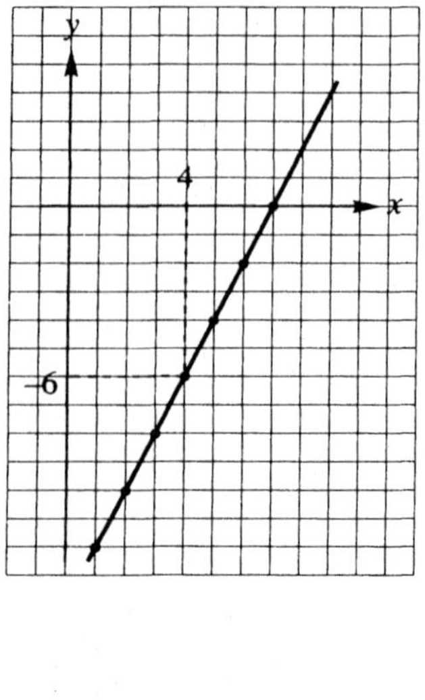

9. $y = 3$

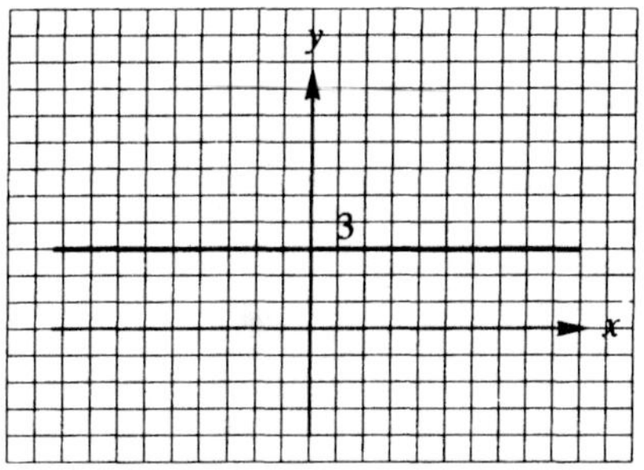

10. $x - 4y = 12$

$$y = \frac{1}{4}x - 3$$

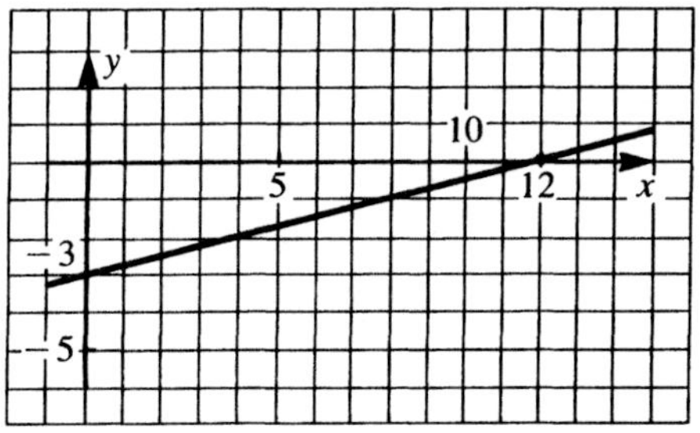

11. (2, 9) and (5, 4)

$$m = \frac{4 - 9}{5 - 2} = -\frac{5}{3}$$

$$y - 9 = -\frac{5}{3}(x - 2) \text{ or } y - 4 = -\frac{5}{3}(x - 5)$$

12. For $3x + 4y = -17$, $m = -\frac{3}{4}$.

$\therefore$ Eqn. is $y = -\frac{3}{4}x + 5$.

13. For $y = 0.25x + 11$, $m = 0.25$

$\therefore$ slope of perp. line is -4.

$\therefore$ Eqn. is $y = -4x$.

14. $x = 5$

15. Not a function

Example:

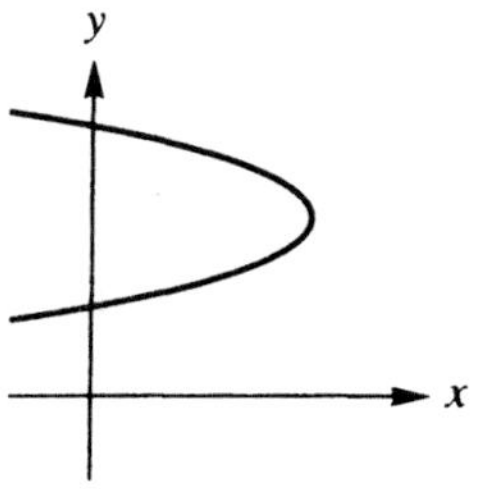

16. Example: -12

17.

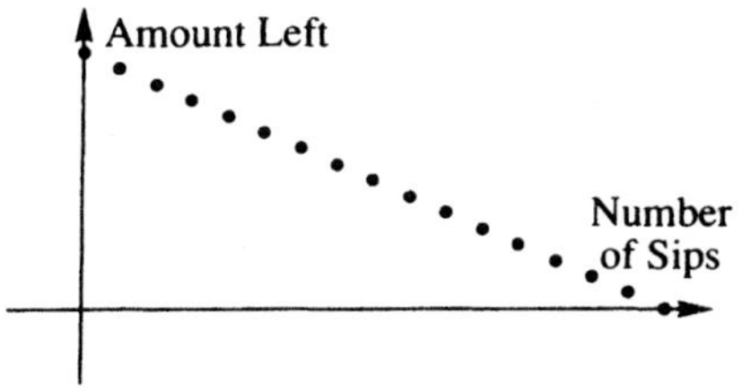

TEST 4 CHAPTER 3 Section 3-5

1. Example:

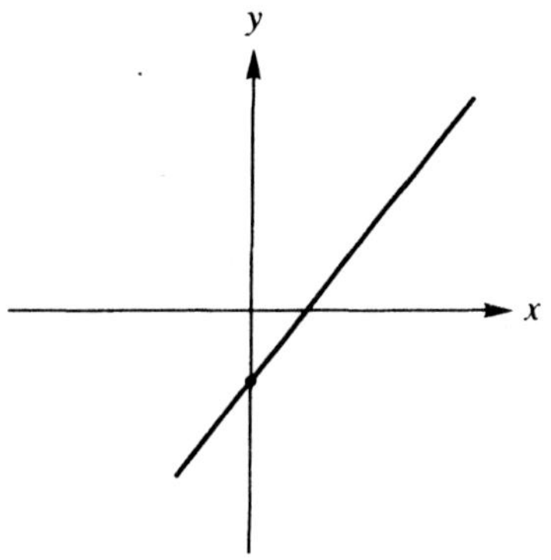

2. Example:

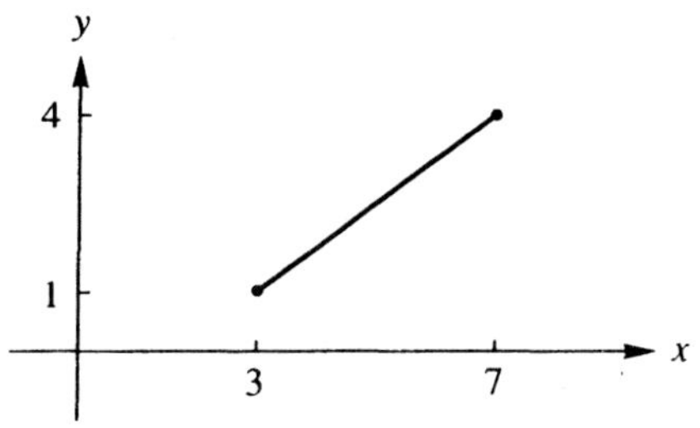

3.

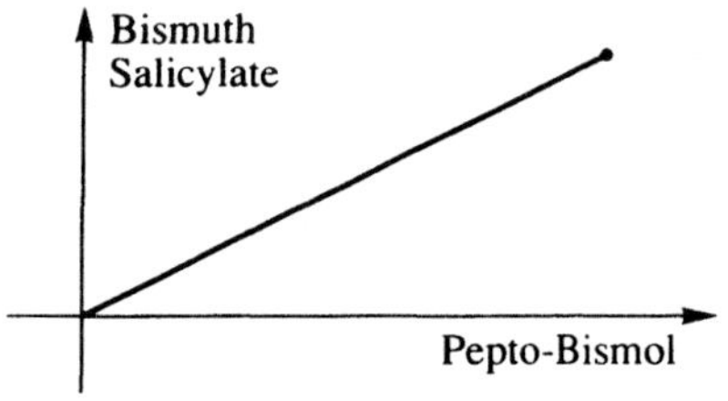

4. *Reading Problem*

Let p = no. of pages left.

Let t = no. of minutes.

(10, 35), (50, 5)

$$m = \frac{5 - 35}{50 - 10} = -\frac{3}{4}$$

$$p - 35 = -\frac{3}{4}(t - 10)$$

$$p = -\frac{3}{4}t + 42\frac{1}{2}$$

5. *Donuts Problem*

a. Let p = no. of dollars.

Let n = no. of donuts.

(5, 1.15), (11, 2.35)

$$m = \frac{2.35 - 1.15}{11 - 5} = 0.2$$

$$p - 1.15 = 0.2(n - 5)$$

$$p = 0.2n + 0.15$$

b. $p = 0.2(3) + 0.15 = 0.75$

75 cents

c. $0.2n + 0.15 = 3.15$
$0.2n = 3.00$
$n = 15$
15 donuts

d.

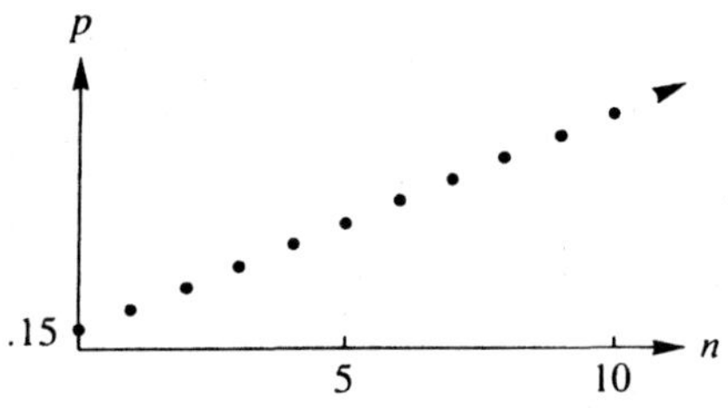

e. The $.15 price-intercept is the fixed charge for the box, service, etc. Slope is the number of dollars per donut.

6. *Elevator Problem*

a. The function is linear since the equation has the form $y = mx + b$.

b. $c = 20(29) + 35$
$c = 615$
615 feet

c. $375 = 20s + 35$
$340 = 20s$
$17 = s$
17 stories

d. The slope is the number of feet per story.

e. c-intercept $= 35$
It represents the number of extra feet of cable in the elevator machinery.

f. Domain is integers greater than or equal to 2.
(No elevator is needed in a one-story building!)

g.

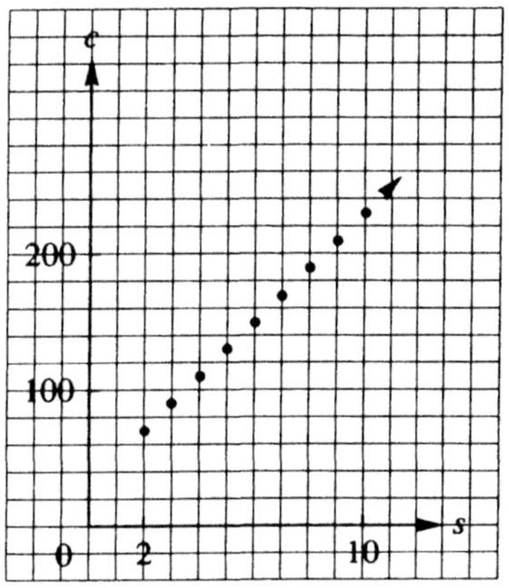

TEST 5 CHAPTER 3

1. $y = mx + b$

2. (3, 5) and (−2, 7)
$m = \dfrac{7 - 5}{-2 - 3} = -\dfrac{2}{5}$
$y - 5 = -\frac{2}{5}(x - 3)$
Alternate forms:
$y - 7 = -\frac{2}{5}(x + 2)$
$y = -\frac{2}{5}x + 6\frac{1}{5}$
$2x + 5y = 31$

3. $y + 1 = -\frac{3}{7}(x - 4)$
Alternate forms:
$y = -\frac{3}{7}x + \frac{5}{7}$
$3x + 7y = 5$

4. (6, 0) and (0, 9)
$m = \dfrac{9 - 0}{0 - 6} = -\dfrac{3}{2}$
$y = -\frac{3}{2}x + 9$
Alternate form:
$3x + 2y = 18$

5. $x = 4$

6. $y = 7$

7. $4x - 8y = 13$
$-8y = -4x + 13$
$y = \frac{1}{2}x - 1\frac{5}{8}$

8. $m = -2$ (The coordinates (−11, 17) are superfluous!)

9. a. $y - 5 = \frac{7}{3}(x + 4)$
point is (−4, 5)

b. $3y - 15 = 7x + 28$
$7x - 3y = -43$

c. If $y = 0$, then $x = -\frac{43}{7}$

10. Closure of {real numbers} under multiplication.

11. $|12x - 71| = 11$
$12x - 71 = 11$ or $12x - 71 = -11$
$12x = 82$ or $12x = 60$
$x = 6\frac{5}{6}$ or $x = 5$
$S = \left\{6\frac{5}{6}, 5\right\}$

12. $(3x + 4)(2x - 7)$
$= 6x^2 - 13x - 28$

13. Let h = no. of hours.
Let d = no. of dollars.
$d = 3.50\,h + 2.00$
If $d = 21.25$, then
$21.25 = 3.50h + 2.00$
$19.25 = 3.50h$
$5.5 = h$
$5\frac{1}{2}$ hours

14.

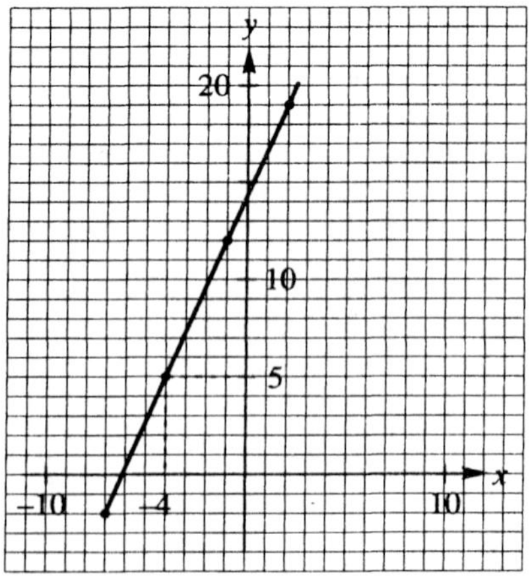

TEST 6 CHAPTER 4 Sections 4-1 to 4-3

1. $3x + 7y = -29$ M4
$4x - 9y = 53$ M3

$$
\begin{aligned}
12x + 28y &= -116 \\
\underline{12x - 27y} &\underline{= 159} \\
55y &= -275 \\
y &= -5 \\
3x - 35 &= -29 \\
3x &= 6 \\
x &= 2
\end{aligned}
$$

$S = \{(2, -5)\}$

2. $5x + 2y = 8$
$3x - 4y = -6$

$$x = \frac{\begin{vmatrix} 8 & 2 \\ -6 & -4 \end{vmatrix}}{\begin{vmatrix} 5 & 2 \\ 3 & -4 \end{vmatrix}} = \frac{-32 + 12}{-20 - 6} = \frac{10}{13}$$

$$y = \frac{\begin{vmatrix} 5 & 8 \\ 3 & -6 \end{vmatrix}}{-26} = \frac{-30 - 24}{-26} = \frac{27}{13}$$

$$S = \left\{\left(\frac{10}{13}, \frac{27}{13}\right)\right\}$$

3. Independent: Example:

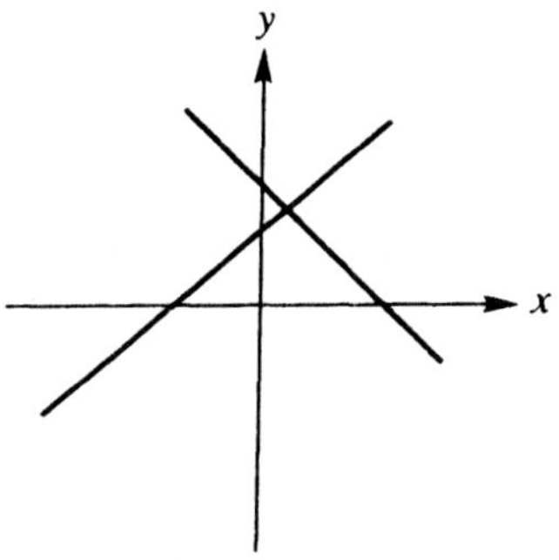

4. $3x - 5y = 15$

$$
\begin{aligned}
\underline{3x - 5y} &\underline{= 20} \\
0 &= -5
\end{aligned}
$$

Inconsistent equations.

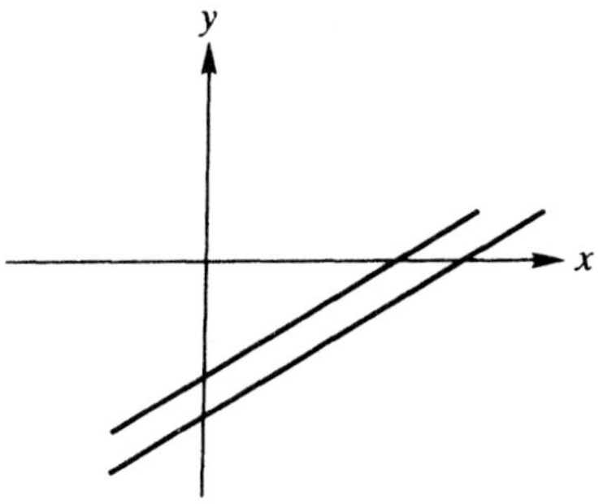

5. $8x + 12y = 36$
$6x + 9y = 27$

$$x = \frac{\begin{vmatrix} 36 & 12 \\ 27 & 9 \end{vmatrix}}{\begin{vmatrix} 8 & 12 \\ 6 & 9 \end{vmatrix}} = \frac{324 - 324}{72 - 72} = \frac{0}{0}$$

Dependent equations.

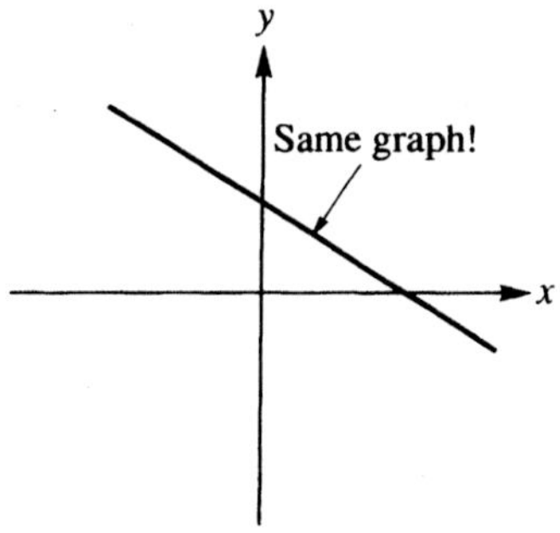

6. Example: $-0.5 = -\frac{1}{2} = \frac{-1}{2}$, which is a ratio of two integers.

7. $y^2 = x + 5, x = 11$:
$y^2 = 11 + 5$
$y^2 = 16$
$y = 4$ or -4
Not a function.
Two values of y for the *same* value of x.

8.

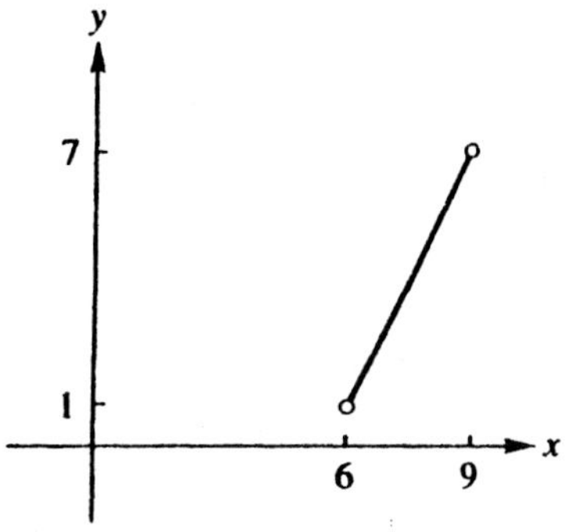

9. *Computer Disk Problem*
Let p = no. of dollars.
Let n = no. of disks.
(10, 15), (30, 40)

$$m = \frac{40 - 15}{30 - 10} = \frac{5}{4}$$

$$p - 15 = \frac{5}{4}(n - 10)$$

$$\underline{\underline{p = \frac{5}{4}n + 2\frac{1}{2}}}$$

10. $y = -\frac{2}{5}x - 3$

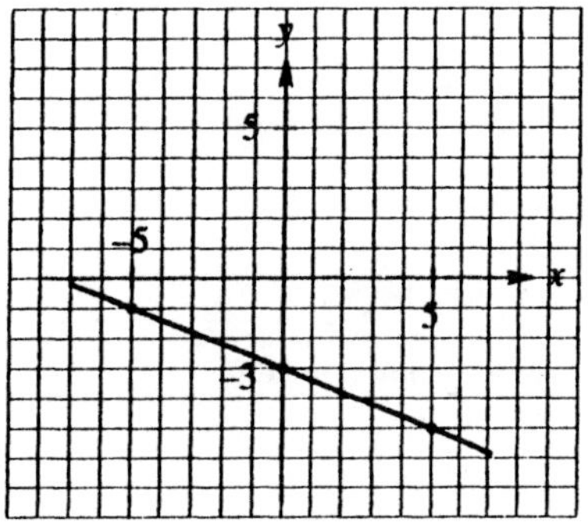

TEST 7 CHAPTER 4 Sections 4-4 to 4-8

1. $\left[\begin{array}{cc|c} 2 & 7 & 8 \\ 5 & -3 & 11 \end{array}\right] \xrightarrow[-① \times 5]{② \times 2} \left[\begin{array}{cc|c} 2 & 7 & 8 \\ 0 & -41 & -18 \end{array}\right] \xrightarrow{\div -41}$

$\left[\begin{array}{cc|c} 2 & 7 & 8 \\ 0 & 1 & 0.439\ldots \end{array}\right] \xrightarrow[-② \times 7]{①} \left[\begin{array}{cc|c} 2 & 0 & 4.926\ldots \\ 0 & 1 & 0.439\ldots \end{array}\right] \xrightarrow{\div 2}$

$\left[\begin{array}{cc|c} 1 & 0 & 2.463\ldots \\ 0 & 1 & 0.439\ldots \end{array}\right]$

$S = \{(2.463\ldots, 0.439\ldots)\}$

2. $3x + 2y - z = 3$ —①

$x + 4y + 2z = 19$ —②

$2x + 3y - 5z = 0$ —③

$2 \times ①: 6x + 4y - 2z = 6$

$②: x + 4y + 2z = 19$

$5x - 4z = -13$ —④

$3 \times ①: 9x + 6y - 3z = 9$

$2 \times ③: 4x + 6y - 10z = 0$

$5x + 7z = 9$ —⑤

$④: 5x - 4z = -13$

$11z = 22$

$z = 2$

$5x + 14 = 9$

$5x = -5$

$x = -1$

$-1 + 4y + 4 = 19$

$4y = 16$

$y = 4$

$S = \{(-1, 4, 2)\}$

3. $\left[\begin{array}{ccc|c} 3 & 2 & -1 & 3 \\ 1 & 4 & 2 & 19 \\ 2 & 3 & -5 & 0 \end{array}\right]$

which transforms to

$\left[\begin{array}{ccc|c} 1 & 0 & 0 & -1 \\ 0 & 1 & 0 & 4 \\ 0 & 0 & 1 & 2 \end{array}\right]$

$\therefore S = \{(-1, 4, 2)\}$

4. $y = \dfrac{\begin{vmatrix} 3 & 3 & -1 \\ 1 & 19 & 2 \\ 2 & 0 & -5 \end{vmatrix}}{\begin{vmatrix} 3 & 2 & -1 \\ 1 & 4 & 2 \\ 2 & 3 & -5 \end{vmatrix}}$

$= \dfrac{3\begin{vmatrix} 19 & 2 \\ 0 & -5 \end{vmatrix} - 3\begin{vmatrix} 1 & 2 \\ 2 & -5 \end{vmatrix} - 1\begin{vmatrix} 1 & 19 \\ 2 & 0 \end{vmatrix}}{3\begin{vmatrix} 4 & 2 \\ 3 & -5 \end{vmatrix} - 2\begin{vmatrix} 1 & 2 \\ 2 & -5 \end{vmatrix} - 1\begin{vmatrix} 1 & 4 \\ 2 & 3 \end{vmatrix}}$

$= \dfrac{-285 + 27 + 38}{-78 + 18 + 5}$

$= \dfrac{-220}{-55} = \underline{\underline{4}}$, which checks.

5. $|4x - 25| > 13$, $x \in$ {natural numbers}

$4x - 25 > 13$ or $4x - 25 < -13$

$x > 9\frac{1}{2}$ or $x < 3$

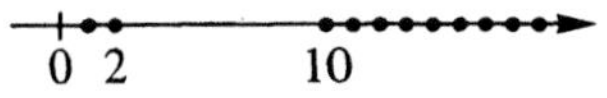

6. $g(x) = 3x - 4$, $h(x) = 2x^3$

$g(3) = 3 \cdot 3 - 4 = \underline{\underline{5}}$

$h(5) = 2 \cdot 5^3 = \underline{\underline{250}}$

$h(g(1)) = h(-1) = \underline{\underline{-2}}$

$g(g(8)) = g(20) = \underline{\underline{56}}$

7. *Swimming Problem*

x = no. of seconds

$f(x)$ = Adolph's no. of m from beach.

$g(x)$ = Sheila's no. of m from beach.

a. $\underline{\underline{f(x) = 130 - 3.2x}}$

b. $f(0) = \underline{\underline{130}}$

$f(20) = \underline{\underline{66}}$

c. $\underline{\underline{g(x) = 80 - 1.7x}}$

d. $g(0) = \underline{\underline{80}}$

$g(20) = \underline{\underline{46}}$

e. $f(x) = g(x)$

$130 - 3.2x = 80 - 1.7x$

$50 = 1.5x$

$33.3\ldots = x$

<u>About 33 sec.</u>

f. $g(33.3\ldots) = 23.3\ldots$

<u>No</u>. Sheila is still 23 m from the beach when Adolph catches up.

g.

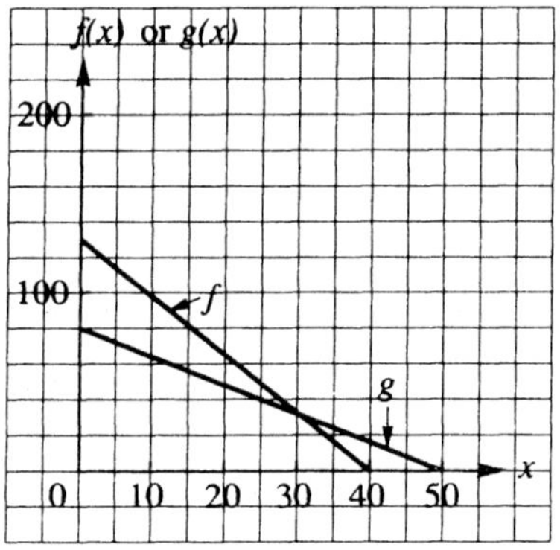

8.

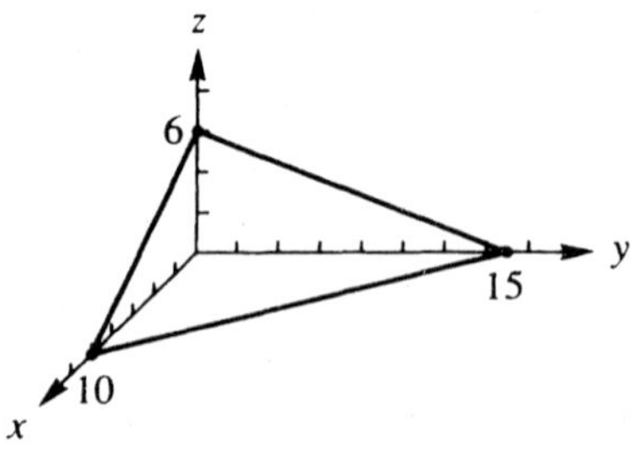

(*Note*: You may have students omit either Problem 3 or Problem 4 depending on time.)

TEST 8 CHAPTER 4 Sections 4-10 and 4-11

1. $f(x) = 3x^2 - 5x + 7$

$f(6) = 108 - 30 + 7$

$= \underline{\underline{85}}$

$f(-4) = 48 + 20 + 7$
$\quad = \underline{\underline{75}}$

2. $3x - 2y \geq 12 \longrightarrow y \leq \frac{3}{2}x - 6$

$y < \frac{1}{5}x + 2$

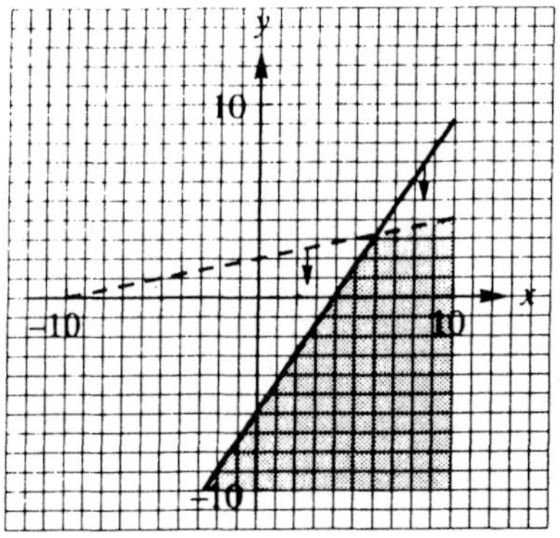

3. Let x = no. of shirts.
Let y = no. of dresses.
i. $x + y \geq 10 \longrightarrow y \geq -x + 10$
ii. $5 \leq x \leq 27$
iii. $y < 2x$ and $y \geq \frac{1}{6}x$
iv. $4x + 3y \leq 135 \longrightarrow y \leq -\frac{4}{3}x + 45$

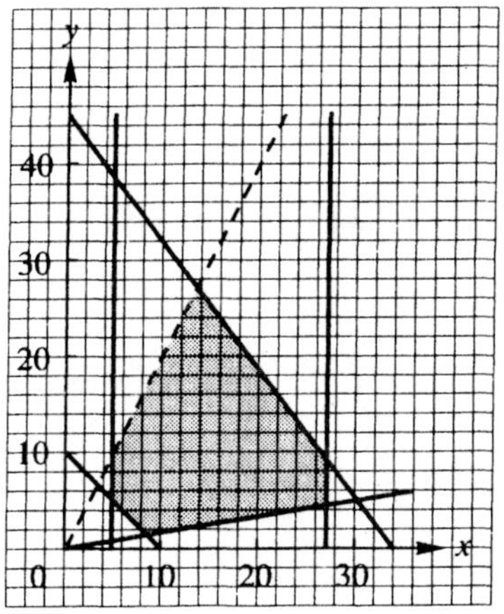

4. a. P = no. of \$ profit.
x = no. of large machines.
y = no. of small machines.
$\underline{\underline{P = 1000x + 500y}}$

b. $1000x + 500y \geq 15{,}000$
$2x + y \geq 30$
$y \geq -2x + 30$

c. Optimum point is (15,14)
<u>15 large machines, 14 small.</u>
$P = 1000(15) + 500(14)$
$\quad = 22{,}000$
$\underline{\underline{\$22{,}000}}$

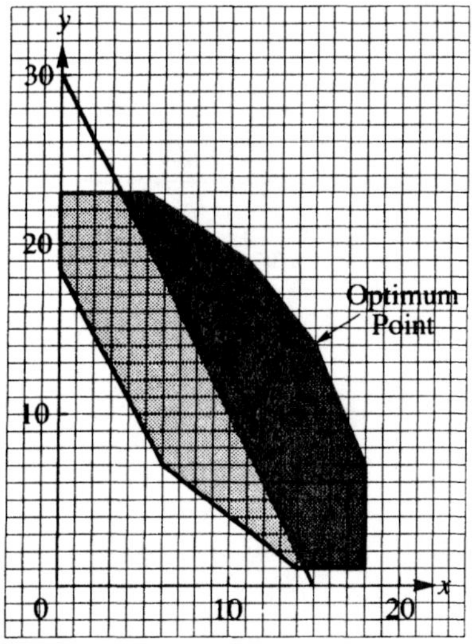

TEST 9 CHAPTER 4

1. $3x - 7y = 13$ —①
$2x + 9y = -4$ —②

$$x = \frac{\begin{vmatrix} 13 & -7 \\ -4 & 9 \end{vmatrix}}{\begin{vmatrix} 3 & -7 \\ 2 & 9 \end{vmatrix}} = \frac{117 - 28}{27 + 14} = \frac{89}{41}$$

$$y = \frac{\begin{vmatrix} 3 & 13 \\ 2 & -4 \end{vmatrix}}{41} = \frac{-12 - 26}{41} = -\frac{38}{41}$$

$$S = \left\{\left(\frac{89}{41}, -\frac{38}{41}\right)\right\} = \{(2.17\ldots, 0.92\ldots)\}$$

2. $5x - 4y = -47$ —①
$7x + 6y = 27$ —②
$3 \times$ ①: $15x - 12y = -141$
$2 \times$ ②: $\underline{14x + 12y = 54}$
$29x = -87$
$x = -3$

$-21 + 6y = 27$
$6y = 48$
$y = 8$
$S = \{(-3, 8)\}$

3. $\left[\begin{array}{cc|c} 5 & 2 & 8 \\ 3 & 1 & 6 \end{array}\right]$

$\left[\begin{array}{cc|c} 5 & 2 & 8 \\ 0 & 1 & -6 \end{array}\right] \leftarrow 3 \times ① - 5 \times ②$

$\left[\begin{array}{cc|c} 5 & 0 & 20 \\ 0 & 1 & -6 \end{array}\right] \leftarrow ① - 2 \times ②$

$\left[\begin{array}{cc|c} 1 & 0 & 4 \\ 0 & 1 & -6 \end{array}\right] \leftarrow ① \div 5$

$S = \{(4, -6)\}$

4. $3x - 4y = 9$ —①
$6x - 8y = 1$ —②
$2 \times$ ①: $6x - 8y = 18$
②: $\underline{6x - 8y = 1}$
$0 = 17$
Inconsistent equations.
$S = \varnothing$

5. $$\begin{aligned} 2x + 3y - z &= -1 \text{ — ①}\\ -x + 5y + 3z &= -10 \text{ — ②}\\ 3x - y - 6z &= 5 \text{ — ③}\end{aligned}$$

$$\begin{aligned} ①:\quad 2x + 3y - z &= -1\\ 2 \times ②:\quad -2x + 10y + 6z &= -20\\ \hline 13y + 5z &= -21 \text{ — ④}\end{aligned}$$

$$\begin{aligned} 3 \times ②:\quad -3x + 15y + 9z &= -30\\ ③:\quad 3x - y - 6z &= 5\\ \hline 14y + 3z &= -25 \text{ — ⑤}\end{aligned}$$

$$\begin{aligned} 3 \times ④:\quad 39y + 15z &= -63\\ 5 \times ⑤:\quad 70y + 15z &= -125\\ \hline -31y &= 62\\ y &= -2\end{aligned}$$

$$\begin{aligned} -26 + 5z &= -21\\ z &= 1\\ 2x - 6 - 1 &= -1\\ x &= 3\end{aligned}$$

$S = \{(3, -2, 1)\}$

(*Note:* This is the Example in Section 4-6.)

6. $$\left[\begin{array}{rrr|r} 2 & 3 & -1 & -1\\ -1 & 5 & 3 & -10\\ 3 & -1 & -6 & 5\end{array}\right] \begin{array}{c} \\ \rightarrow \\ \rightarrow \end{array} \left[\begin{array}{rrr|r} 2 & 3 & -1 & -1\\ 0 & 13 & 5 & -21\\ 0 & -11 & -9 & 13\end{array}\right] \rightarrow$$

$$\left[\begin{array}{rrr|r} 2 & 3 & -1 & -1\\ 0 & 13 & 5 & -21\\ 0 & 0 & -62 & -62\end{array}\right] \rightarrow \left[\begin{array}{rrr|r} 2 & 3 & -1 & -1\\ 0 & 13 & 5 & -21\\ 0 & 0 & 1 & 1\end{array}\right]$$

$\therefore \underline{\underline{z = 1}}$

7. $$z = \frac{\begin{vmatrix} 2 & 3 & -1\\ -1 & 5 & -10\\ 3 & -1 & 5\end{vmatrix}}{\begin{vmatrix} 2 & 3 & -1\\ -1 & 5 & 3\\ 3 & -1 & -6\end{vmatrix}} = \frac{-31}{-31} = \underline{\underline{1}}$$

8. $f(x) = 3x - 7;\ g(x) = 4x + 5$

Let $f(x) = g(x)$.

$$\begin{aligned} \therefore 3x - 7 &= 4x + 5\\ -x &= 12\\ x &= -12\end{aligned}$$

$f(-12) = -43,\ g(-12) = -43\ \checkmark$

Point is $\underline{\underline{(-12, -43)}}$

9.

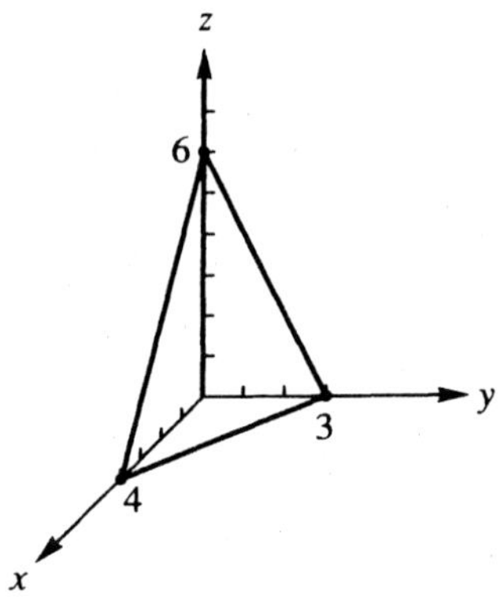

10. i. $x + y \le 10$ ii. $y < 2x$

iii. $y \ge \frac{1}{3}x + 2$

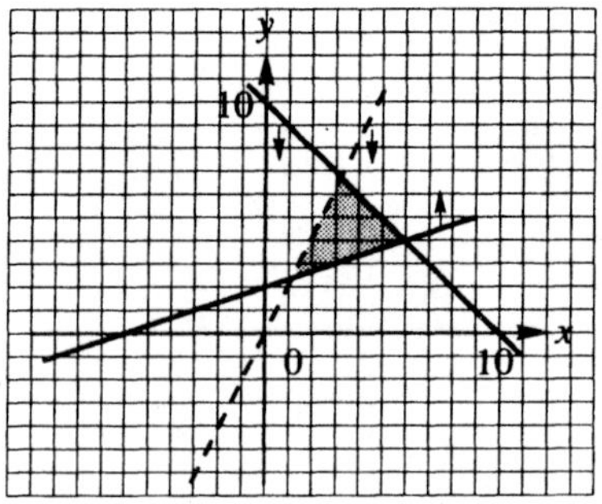

TEST 10 CHAPTER 5 Sections 5-1 to 5-5

1. $f(x) = 5x^2 - 8x + 13$

a. $f(3) = 45 - 24 + 13$
$= \underline{\underline{34}}$

b. $f(-7) = 245 + 56 + 13$
$= \underline{\underline{314}}$

2. $5x^2 - 8x - 3 = 0$

$$x = \frac{8 \pm \sqrt{64 - 4(5)(-3)}}{10}$$

$$x = \frac{8 \pm \sqrt{124}}{10}$$

$S = \{1.913\ldots,\ -0.313\ldots\}$

3.

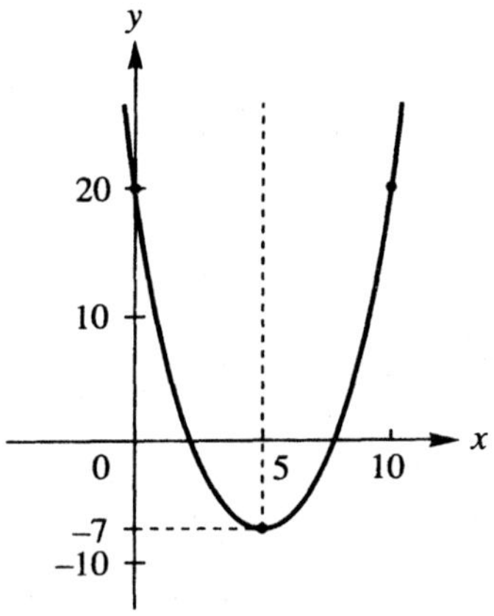

4. Example:

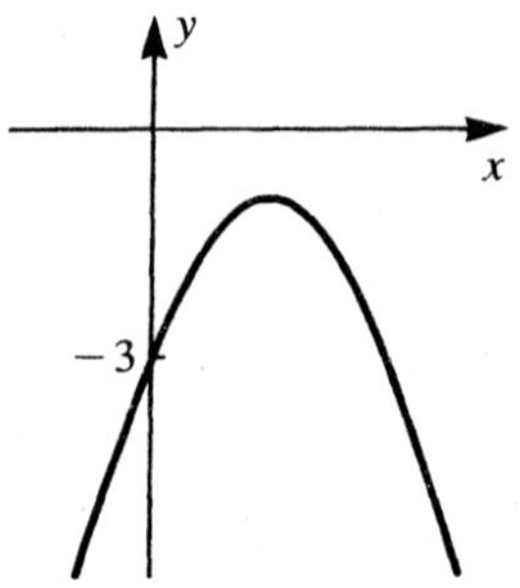

5. Example:

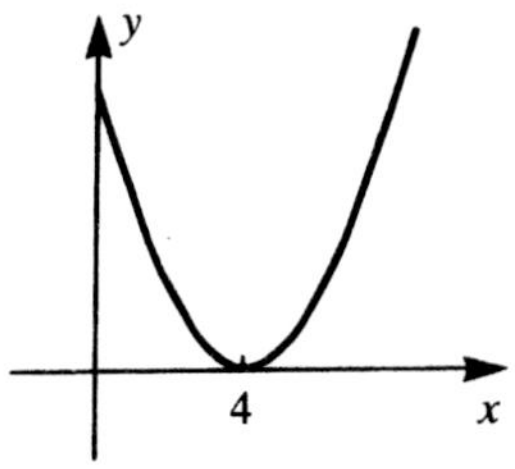

6. $y - 8 = -3(x + 4)^2$
Vertex: $(-4, 8)$
$x = 0: y - 8 = -3(4^2)$
$y = -40$

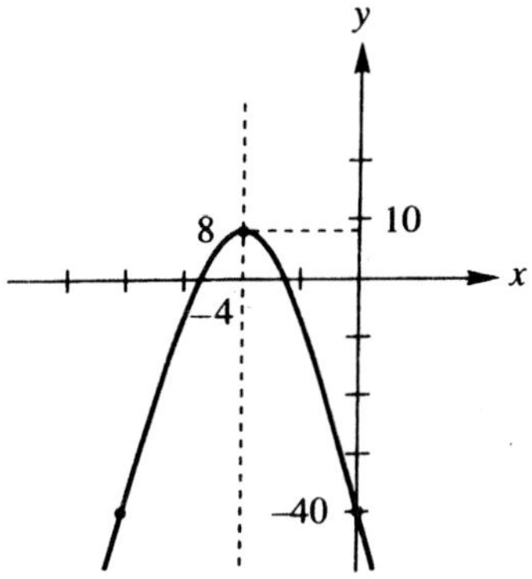

7. $y = 7x^2 + 42x - 13$
$y + 13 + 63 = 7(x^2 + 6x + 9)$
$y + 76 = 7(x + 3)^2$

8. $y = x^2 - 13x - 3$
If $y = 0$, then

$$x = \frac{13 \pm \sqrt{169 - 4(1)(-3)}}{2}$$

$$x = \frac{13 \pm \sqrt{181}}{2}$$

Intercepts are: 13.226... , − 0.226...

9. $3x^2 + 7x + 5 = 0$
$b^2 - 4ac = 49 - 4(3)(5)$
$= -11$

No real solutions.

11. $(-1, -2), (3, 10)$
$$m = \frac{10 + 2}{3 + 1} = 3$$
$y - 10 = 3(x - 3)$
$y = 3x + 1$

10. a. $x^2 - 12x + 52 = 0$

$$x = \frac{12 \pm \sqrt{-64}}{2}$$

$x = 6 \pm 4i$
$S = \{6 + 4i, 6 - 4i\}$

b.

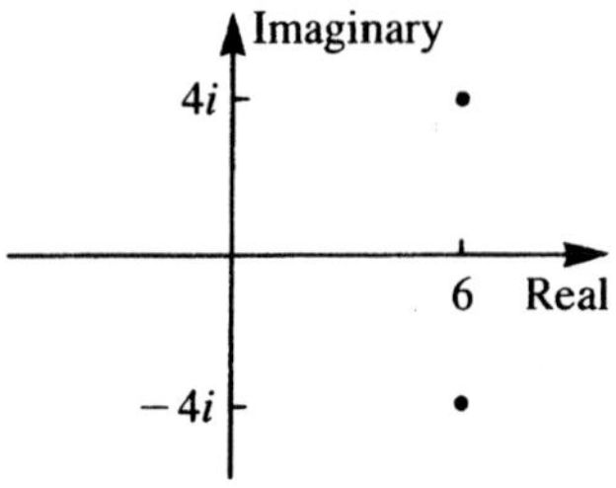

c. Conjugates.

d. Check $6 + 4i$:
$(6 + 4i)^2 - 12(6 + 4i) + 52 \stackrel{?}{=} 0$
$36 + 48i + 16i^2 - 72 - 48i + 52 \stackrel{?}{=} 0$
$36 - 16 - 72 + 52 \stackrel{?}{=} 0$
$0 = 0$

12. $(5x + 11)^2 = 25x^2 + 110x + 121$

13. $x^2 - 3x - 10 = (x - 5)(x + 2)$

TEST 11 CHAPTER 5 Sections 5-7 and 5-8

Note: Problems 1 and 2 are the same as Examples 1 and 3, respectively, in Section 5-6.

1. $(-2, -11), (4, 13), (6, 29)$
$4a - 2b + c = -11$
$16a + 4b + c = 13$
$36a + 6b + c = 29$
$12a + 6b = 24 \longrightarrow 2a + b = 4$
$32a + 8b = 40 \longrightarrow 4a + b = 5$
$2a = 1$
$a = \frac{1}{2}$

$1 + b = 4$
$b = 3$
$2 - 6 + c = -11$
$c = -7$
$\therefore y = \frac{1}{2}x^2 + 3x - 7$

2. Vertex: $(2, -5)$; point: $(3, 1)$
$y + 5 = a(x - 2)^2$
$1 + 5 = a(3 - 2)^2$
$6 = a$
$\therefore y + 5 = 6(x - 2)^2$
$y = 6x^2 - 24x + 19$

3. *Pizza Problem*

a. Let d = no. of inches.
Let p = no. of dollars.
(10, 4.30), (12, 5.88), (14, 7.78)
$100a + 10b + c = 4.30$
$144a + 12b + c = 5.88$
$196a + 14b + c = 7.78$
$44a + 2b = 1.58$
$52a + 2b = 1.90$
$8a = 0.32$
$a = 0.04$
$1.76 + 2b = 1.58$
$2b = -0.18$
$b = -0.09$

$4 - 0.9 + c = 4.30$

$c = 1.20$

$\therefore \underline{p = 0.04d^2 - 0.09d + 1.20}$

b. $p = 0.04(25) - 0.09(5) + 1.20$

$= 1.75$

$\underline{\$1.75}$

c. $20.00 = 0.04d^2 - 0.09d + 1.20$

$0.04d^2 - 0.09d - 18.80 = 0$

$$d = \frac{0.09 \pm \sqrt{0.0081 - 4(0.04)(-18.80)}}{0.08}$$

$$= \frac{0.09 \pm \sqrt{3.0161}}{0.08}$$

$= 22.83\ldots$ or $-20.58\ldots$

About 23 inches

d. Price-intercept tells that $1.20 goes for cooking, service, and other fixed charges.

4. Let A = no. of in^2 area.

$A = (10 - x)(8 - x)$

$\underline{A = x^2 - 18x + 80}$

5. $17 - 4x > 28$

$-4x > 11$

$x < -2\frac{3}{4}$

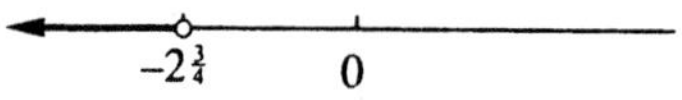

TEST 12 CHAPTER 5

1. $y = 3x^2 + 24x + 43$

a. $y - 43 + 48 = 3(x^2 + 8x + 16)$

$\underline{y + 5 = 3(x + 4)^2}$

b. Vertex: $\underline{(-4, -5)}$

c. $3x^2 + 24x + 43 = 0$

$$x = \frac{-24 \pm \sqrt{576 - 4(3)(43)}}{6}$$

$$x = \frac{-24 \pm \sqrt{60}}{6}$$

$x = \underline{-2.709\ldots \text{ or } -5.290\ldots}$

d.

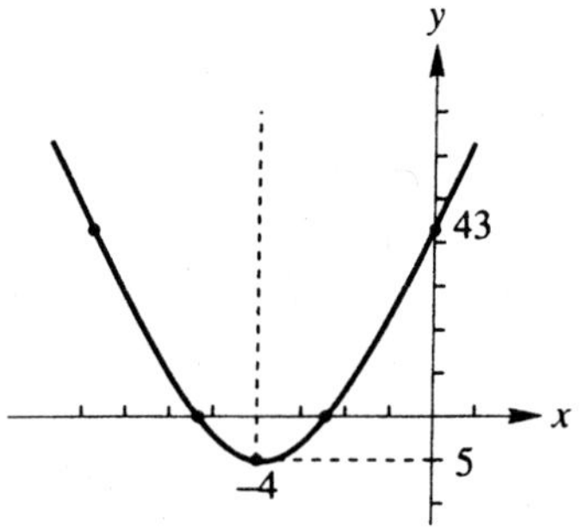

2. $x = \dfrac{5 \pm \sqrt{25 - 4(7)(-3)}}{14}$

$x = \dfrac{5 \pm \sqrt{109}}{14}$

$S = \{1.102\ldots, -0.388\ldots\}$

3. $x^2 + 2x + 26 = 0$

$x = \dfrac{-2 \pm \sqrt{-100}}{2}$

$x = -1 \pm 5i$

$\underline{S = \{-1 + 5i, -1 - 5i\}}$

4. $(5 + 7i)(4 - 3i)$

$= 20 + 13i - 21i^2$

$= \underline{41 + 13i}$

5. $f(x) = 11x^2 + 13x - 28$

$f(x) = -100$:

$11x^2 + 13x - 28 = -100$

$11x^2 + 13x + 72 = 0$

$b^2 - 4ac = 169 - 4(11)(72)$

$= -2999$

$\therefore f(x)$ never equals -100.

6.

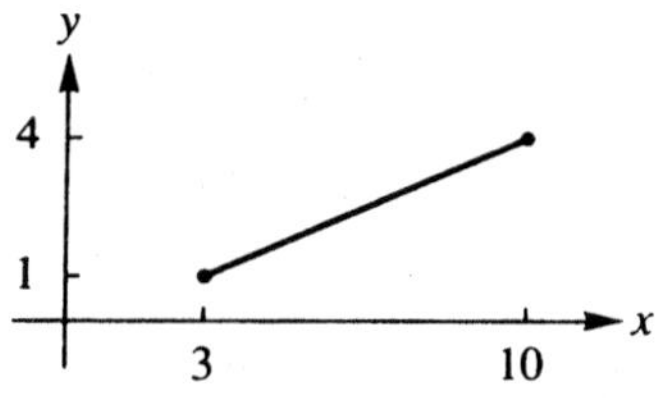

Note: Slope could be negative.

7. *Vertical Motion Problem*

a. Let d = no. of m above ground.

Let t = no. of seconds.

$d = at^2 + bt + c$

(1, 68), (2, 96), (3, 114)

$a + b + c = 68$

$4a + 2b + c = 96$

$9a + 3b + c = 114$

$3a + b = 28$

$5a + b = 18$

$-2a = 10$

$a = -5$

$-15 + b = 28$

$b = 43$

$-5 + 43 + c = 68$

$c = 30$

$\therefore$ equation is $\underline{d = -5t^2 + 43t + 30}$

b. Vertex at $t = -\dfrac{b}{2a} = -\dfrac{43}{-10} = 4.3$

$d = -5(4.3)^2 + 43(4.3) + 30$

$= 122.45$

$\underline{122.45\ m}$

c. When $t = 0$, rock is at the platform. So the platform is $\underline{30\ m}$ high.

d. $\underline{43\ \text{m/sec}}$

e. If $t = 10$,

$d = -500 + 430 + 30$

$= -40$

Well is $\underline{40\ m}$ deep

TEST 13 CHAPTERS 1 through 5

1. $y = mx + b$

2. $y = ax^2 + bx + c$

3. Distributivity

4. Associative axiom for addition

5. Multiplicative identity

6. Symmetry

7. Example: $1 \div 2 = \frac{1}{2}$, and $\frac{1}{2}$ is not an integer.

8. $|3x - 7| \leq 11, x \in \{\text{integers}\}$
$-11 \leq 3x - 7 \leq 11$

$-1\frac{1}{3} \leq x \leq 6$

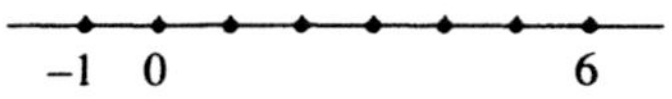

9. $(3x - 7)(x + 4) = 0$
$3x - 7 = 0 \quad \text{or} \quad x + 4 = 0$
$x = \frac{7}{3} \qquad \text{or} \quad x = -4$
$S = \left\{\frac{7}{3}, -4\right\}$

10. $4x^2 - 12x + 13 = 0$
$x = \frac{12 \pm \sqrt{-64}}{8}$
$x = 1.5 \pm i$
$S = \{1.5 + i, 1.5 - i\}$

11. A relation is a set of ordered pairs.

12. Example: $x^3 + 1$

13. Examples: $\frac{x}{y}, \sqrt{x}, |x|$

14. Discriminant $= s^2 - 4rt$

15. Example: $-\frac{1}{2}$

16. $13 - 2 \cdot 12 \div 6 + 5$
$= 13 - 4 + 5$
$= 14$

17. $7 - 3[p - 2(5p - 6)]$
$= 7 - 3[-9p + 12]$
$= 7 + 27p - 36$
$= 27p - 29$

18. $(3x - 4)(2x + 5)$
$= 6x^2 + 7x - 20$

19. $2x^2 - 11x + 12$
$= (2x - 3)(x - 4)$

20. $y = x^2 + 2x - 8$
$h = -\frac{b}{2a} = -\frac{2}{2(1)} = -1$
$k = 1 - 2 - 8 = -9$
Vertex at $(-1, -9)$
y-intercept $= -8$
$x^2 + 2x - 8 = 0$
$(x + 4)(x - 2) = 0$
$x = -4, 2$
x-intercepts are -4 and 2

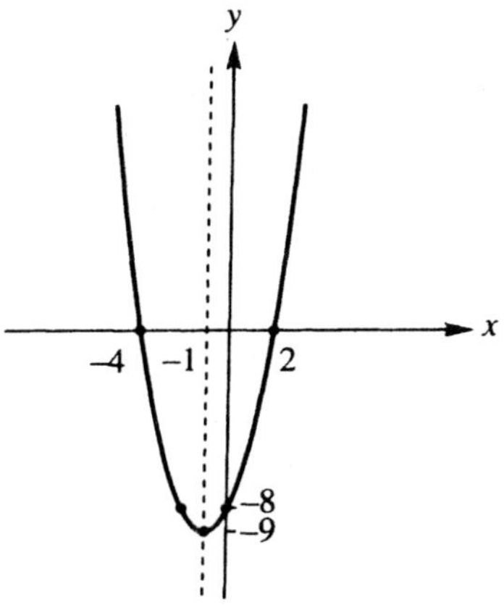

21. $y - 3 = -\frac{5}{2}(x + 4)$

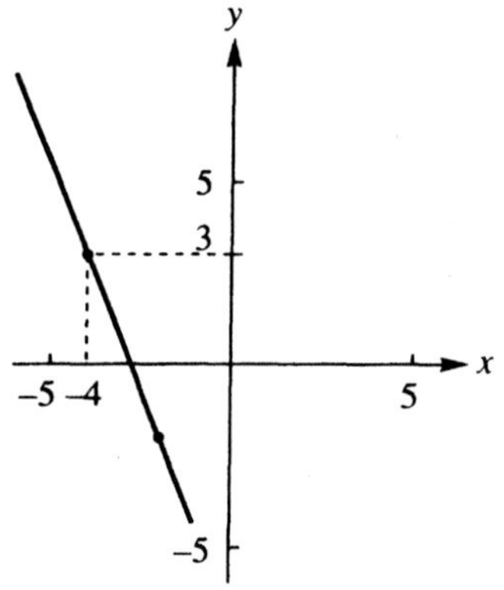

22. $21x + 70y + 30z = 210$

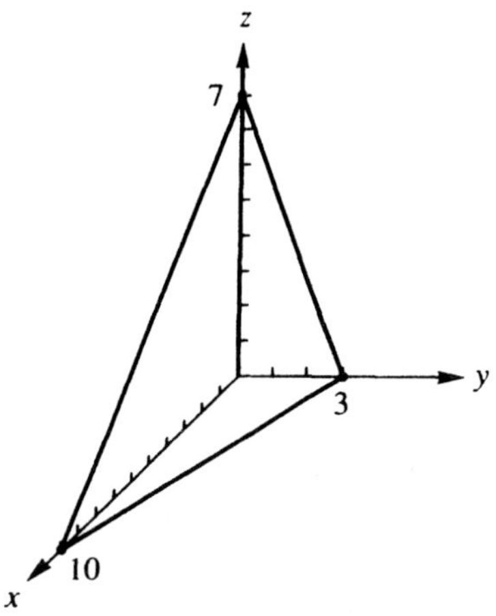

23. $y = \sqrt{x}$

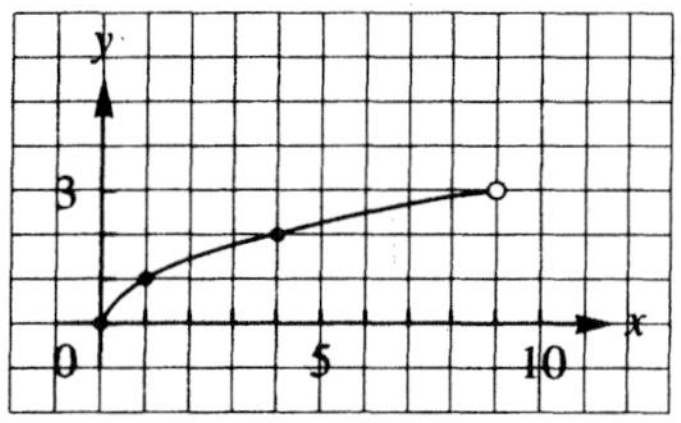

24. Example:

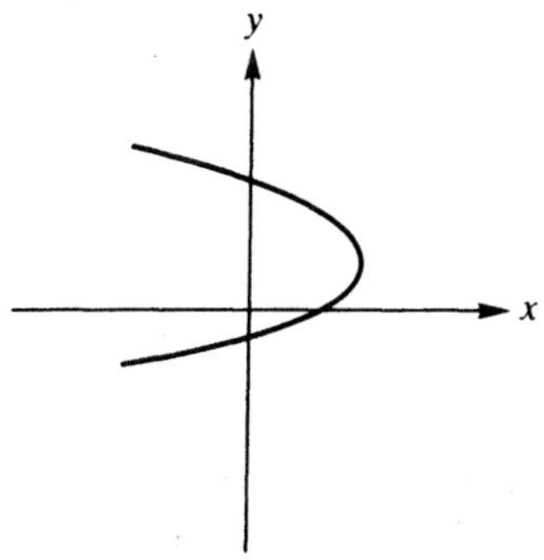

25.

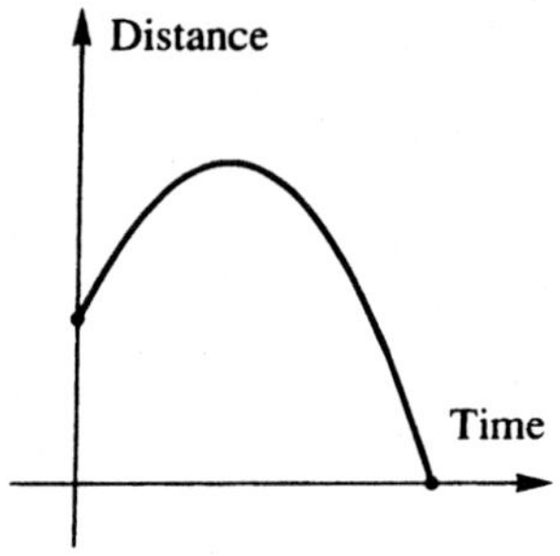

26. $m = 7$

$\therefore y - 5 = 7(x + 3)$

$y - 5 = 7x + 21$

$y = 7x + 26$

27. $y = ax^2 + bx + c$

$$\begin{aligned} a + b + c &= 16 \\ 4a + 2b + c &= 15 \\ 16a + 4b + c &= 1 \end{aligned}$$

$$\begin{aligned} 3a + b &= -1 \longrightarrow 3a + b = -1 \\ 12a + 2b &= -14 \longrightarrow 6a + b = -7 \end{aligned}$$

$$3a = -6$$

$$a = -2$$

$-6 + b = -1$

$b = 5$

$-2 + 5 + c = 16$

$c = 13$

$\therefore$ equation is $y = -2x^2 + 5x + 13$

28. $f(x) = 3x + 7$

$f(-4) = 3(-4) + 7$

$= -5$

29. $g(x) = 5x^2 - 3x + 6$

$g(7) = 5 \cdot 49 - 3 \cdot 7 + 6$

$= 230$

30. $y = 3x^2 - 5x - 2$ and $y = 11$

$3x^2 - 5x - 2 = 11$

$3x^2 - 5x - 13 = 0$

$$x = \frac{5 \pm \sqrt{25 - 4(3)(-13)}}{6}$$

$x = 3.0756\ldots$ or $-1.4089\ldots$

31. $y < -3x + 7$

$2x - 3y \le 12 \longrightarrow y \ge \frac{2}{3}x - 4$

$y \ge -8$

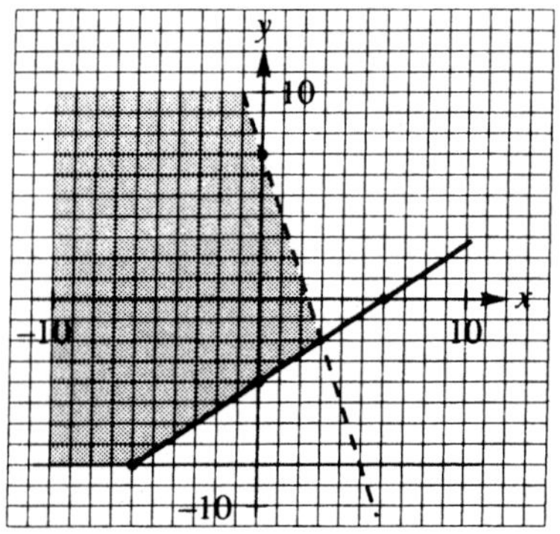

32.
$$\begin{aligned} 3x + 5y + 2z &= -8 \\ 4x - y + 3z &= 7 \\ x + 2y + 5z &= 1 \end{aligned}$$

$$N_z = \begin{vmatrix} 3 & 5 & -8 \\ 4 & -1 & 7 \\ 1 & 2 & 1 \end{vmatrix}$$

$$= 3\begin{vmatrix} -1 & 7 \\ 2 & 1 \end{vmatrix} - 5\begin{vmatrix} 4 & 7 \\ 1 & 1 \end{vmatrix} + (-8)\begin{vmatrix} 4 & -1 \\ 1 & 2 \end{vmatrix}$$

$= -45 + 15 - 72$

$= -102$

33.
$$\begin{aligned} 4x + 7y &= 1 \\ 3x - 5y &= 11 \end{aligned}$$

$$\left[\begin{array}{cc|c} 4 & 7 & 1 \\ 3 & -5 & 11 \end{array}\right]$$

which transforms to

$$\left[\begin{array}{cc|c} 1 & 0 & 2 \\ 0 & 1 & -1 \end{array}\right]$$

$S = \{(2, -1)\}$

34. a. Let x = no. of prints.

Let y = no. of dollars.

(12, 7.33), (15, 8.29)

$$m = \frac{8.29 - 7.33}{15 - 12} = 0.32$$

$y - 7.33 = 0.32(x - 12)$

$y = 0.32x + 3.49$

b. $y = 0.32(36) + 3.49$

$= 15.01$

\$15.01

c. $12.77 = 0.32x + 3.49$

$9.28 = 0.32x$

$29 = x$

29 prints

d. dollars-intercept = 3.49, representing cost to develop the film.

e. Cost per print is \$0.32. The slope tells you.

f.

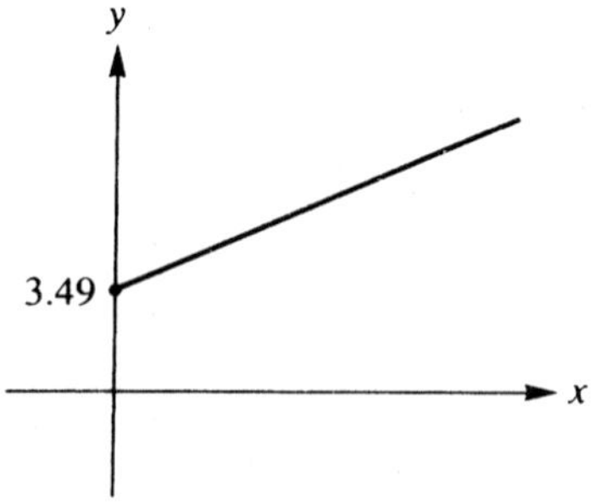

35. a. Let P = no. of dollars profit.

$P = 20x + 30y$

b. $20x + 30y \ge 15{,}000$

$$y \ge -\frac{2}{3}x + 500$$

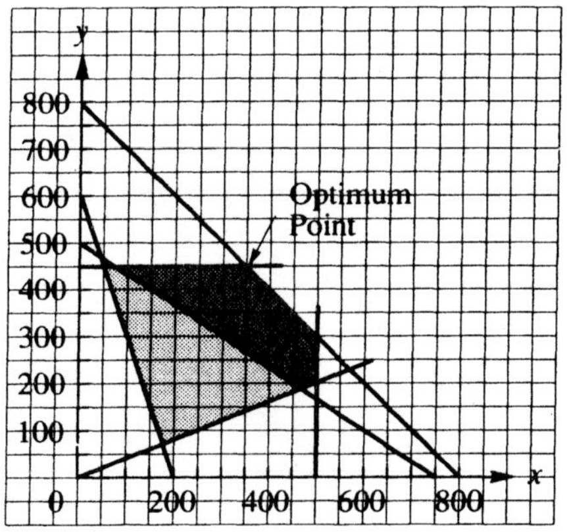

c. Optimum point is (350, 450).
350 plain and 450 deluxe

d. $P = 20(350) + 30(450)$
$= 20{,}500$
$20,500

TEST 14 CHAPTER 6 Sections 6-1 to 6-5

1. $y = a \cdot b^x$

2. $f(x) = 2^x$

x	$f(x)$
-3	1/8
-2	1/4
-1	1/2
0	1
1	2
2	4
3	8

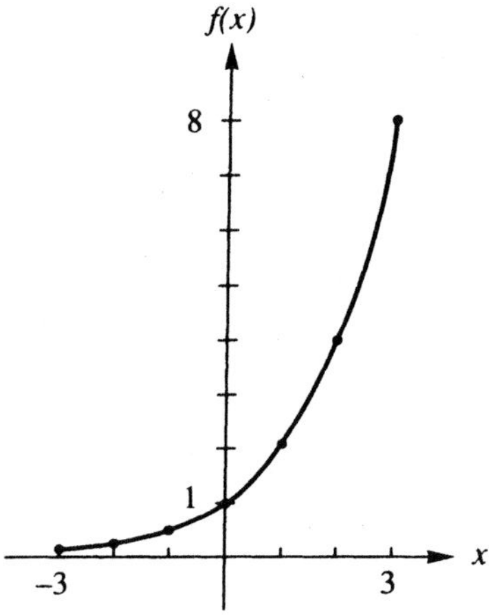

3. $f(x) = 3 \cdot 5^x$
$f(3) = 3 \cdot 5^3 = \underline{\underline{375}}$
$f(-2) = 3 \cdot 5^{-2} = \underline{\underline{\dfrac{3}{25}}}$
$f(0) = 3 \cdot 5^0 = \underline{\underline{3}}$

4. $(ab)^k = \underline{\underline{a^kb^k}}$

5. $\dfrac{x^a}{x^b} = \underline{\underline{x^{a-b}}}$

6. $(y^u)^p = \underline{\underline{y^{up}}}$

7. y, exponent
x, base
x^y, power

8. $2^{3.7} = \underline{\underline{12.99603\ldots}}$

9. $0.8^{-\frac{3}{7}} = \underline{\underline{1.100355\ldots}}$

10. 2^3 is 8 and 2^4 is 16.
So $2^{3.7}$ is between 8 and 16, which the answer to Problem 8 is.

11. $8^{-\frac{2}{3}} = \dfrac{1}{8^{2/3}} = \dfrac{1}{(\sqrt[3]{8})^2} = \dfrac{1}{2^2} = \dfrac{1}{4} = 0.25$, Q.E.D.

12. $5^x = 625$, so $\underline{\underline{x = 4}}$.

13. $100^\circ = \underline{\underline{1}}$.

14. $(4x^6y^{10})^3 = \underline{\underline{64x^{18}y^{30}}}$

15. $\sqrt[6]{(11^{7.5})(11^{-5.1})} = (11^{2.4})^{\frac{1}{6}}$
$= 11^{0.4}$

16. $(6x^{\frac{2}{3}})(5x^{-\frac{1}{5}}y^7) = \underline{\underline{30x^{\frac{7}{15}}y^7}}$

17. $\left(\dfrac{x^5}{y^6}\right)^7\left(\dfrac{y^8}{x^9}\right)^{10}$
$= \dfrac{x^{35}}{y^{42}} \cdot \dfrac{y^{80}}{x^{90}}$
$= \underline{\underline{x^{-55}y^{38}}}$

18. $\dfrac{3 \cdot 2001^{501}}{5 \cdot 2001^{500}}$
$= (0.6)(2001)$
$= \underline{\underline{1200.6}}$

19. $y = ax^2 + bx + c$

20. $14 - 3x > 29$
$-3x > 15$
$x < -5$

TEST 15 CHAPTER 6 Sections 6-6 to 6-12

1. $(8.7 \times 10^{38})(9.324 \times 10^{-13})$
$= 8.11188 \times 10^{26}$
$\approx \underline{\underline{8.1 \times 10^{26}}}$

2. $\dfrac{3 \times 10^{871}}{5 \times 10^{265}}$
$= 0.6 \times 10^{606}$
$= \underline{\underline{6 \times 10^{605}}}$

3. $(3 \times 10^{-7})^4$
$= 81 \times 10^{-28}$
$= \underline{\underline{8.1 \times 10^{-27}}}$

4. $\dfrac{1}{2.5 \times 10^9}$
$= \underline{\underline{4 \times 10^{-10}}}$

5. a. Mantissa.
b. Characteristic.

6. $4.86^{5x} = 3267$
$\log 4.86^{5x} = \log 3267$
$5x \cdot \log 4.86 = \log 3267$
$x = \dfrac{\log 3267}{5 \cdot \log 4.86}$
$x = \underline{\underline{1.0235\ldots}}$

Check: $4.86^{5(1.0235\ldots)}$
$= 3267$ ✓

7. $\log r^p = \underline{\underline{p \cdot \log r}}$

8. $\log\left(\dfrac{57}{19}\right) = \log 3 = 0.4771\ldots$
$\log 57 - \log 19$ — same.
$= 1.7558\ldots - 1.2787\ldots = 0.4771\ldots$

9. Prove that $\log_b (rk) = \log_b r + \log_b k$.

Proof:

Let $x = \log_b r$ and $y = \log_b k$.

$\therefore b^x = r$ and $b^y = k$.

$\therefore rk = b^x \cdot b^y = b^{x+y}$

$\therefore \log_b (rk) = x + y$

$\therefore \log_b (rk) = \log_b r + \log_b k$, Q.E.D.

10. $\log_3 24 + \log_3 17 - \log_3 51$

$= \log_3 \dfrac{24 \times 17}{51}$

$= \underline{\underline{\log_3 8}}$

11. $\log_b 25$

$= 2(1.234)$

$= \underline{\underline{2.468}}$

12. $\log_3 x = 5$

$3^5 = x$

$x = \underline{\underline{243}}$

13. $\log_{32} 8 = x$

$32^x = 8$

$2^{5x} = 2^3$

$5x = 3$

$x = \underline{\underline{0.6}}$

14. $\log_x 64 = \dfrac{2}{3}$

$x^{\frac{2}{3}} = 64$

$x = 64^{\frac{3}{2}}$

$x = \underline{\underline{512}}$

15. Let $x = \log_7 451.9$.

Then $7^x = 451.9$

$\log 7^x = \log 451.9$

$x \cdot \log 7 = \log 451.9$

$x = \dfrac{\log 451.9}{\log 7}$

$x = \underline{\underline{3.14169\ldots}}$

(*Note*: Students who have learned the change-of-base theorem may start at the next-to-last step.)

16. a. $f(x) = 7x - 28$

$x = 7 \cdot f^{-1}(x) - 28$

$x + 28 = 7 \cdot f^{-1}(x)$

$\underline{\underline{f^{-1}(x) = \frac{1}{7}x + 4}}$

b. $f^{-1}(35) = \underline{\underline{9}}$

c. $f(11) = \underline{\underline{49}}$

d. $f(f^{-1}(x))$

$= 7\left(\frac{1}{7}x + 4\right) - 28$

$= x + 28 - 28$

$= x$, Q.E.D.

17. a.

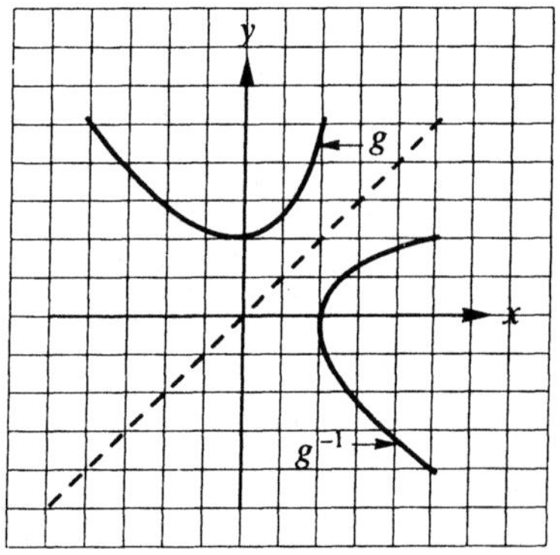

(*Note*: Some students may be clever and fold the question sheet along the dotted line to draw the graph!)

b. g^{-1} is not a function since there are two different values of y for some values of x.

TEST 16 CHAPTER 6 Sections 6-13 and 6-14

1. $\log (3 \cdot 4) = \underline{\underline{\log 3 + \log 4}}$

2. $\log 3^4 = \underline{\underline{4 \cdot \log 3}}$

3. $\log \left(\dfrac{3}{4}\right) = \underline{\underline{\log 3 - \log 4}}$

4. $3.7^{4x} = 9271$

$4x \cdot \log 3.7 = \log 9271$

$x = \dfrac{\log 9271}{4 \cdot \log 3.7}$

$x = 1.745474\ldots$

$S = \{1.745\ldots\}$

5.

x	$f(x)$
5	20
9	14
13	$\underline{\underline{9.8}}$
17	$\underline{\underline{6.86}}$
21	$\underline{\underline{4.802}}$

(+4 for each step in x; $\times .7$ for each step in $f(x)$)

6.

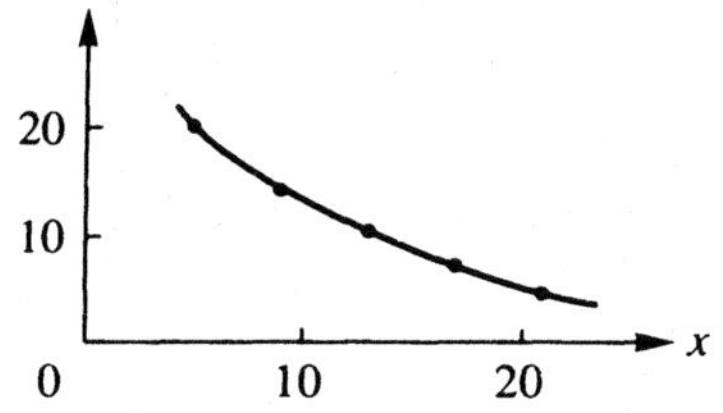

7.

x	y
3	72
5	200
7	392

(+2, +2 in x; $\times 2.777\ldots$ and $\times 1.96$ in y — not the same!)

The data do not have the add-multiply property. Therefore, the function is not exponential.

8. Assume $y = kx^2$

Using (3, 72)

$72 = k \cdot 3^2$

$8 = k$

$\therefore y = 8x^2$

Using this equation for the other two x-values,

$y = 8 \times 5^2 = 200$ ✓

$y = 8 \times 7^2 = 392$ ✓

Therefore, a direct square variation function fits the data.

Alternate method:

$\dfrac{72}{3^2} = 8, \dfrac{200}{5^2} = 8, \dfrac{392}{7^2} = 8.$

$\therefore y$ is directly proportional to the square of x.

9. $g(x) = a \cdot b^t$

(0, 37), (7, 5940)

$a = 37$

$5940 = 37b^7$

$\left(\dfrac{5940}{37}\right)^{\frac{1}{7}} = b$

$b = 2.065\ldots$

$\therefore g(x) = 37 \times 2.065^x\ldots$

10. $h(t) = 17.4 \times 1.23^t$

$h(13) = 17.4 \times 1.23^{13}$

$= \underline{\underline{256.634\ldots}}$

11. $z(t) = 60 \times 1.03^t$
$144 = 60 \times 1.03^t$
$\log\left(\frac{144}{60}\right) = t \cdot \log 1.03$
$\frac{\log\left(\frac{144}{60}\right)}{\log 1.03} = t$
$\underline{\underline{t = 29.6178\ldots}}$

12. *Water Hyacinth Problem*
Let $P(t)$ = no. of plants.
Let t = no. of days after Tues.
$P(t) = C \times b^t$
(0, 150), (3, 240)
$C = 150$
$240 = 150 \times b^3$
$\frac{240}{150} = b^3$
$b = \left(\frac{240}{150}\right)^{\frac{1}{3}}$
$= 1.169607\ldots$
$\therefore P(t) = 150 \times (1.169607\ldots)^t$
Method 1: Find $P(20)$.
$P(20) = 150 \times 1.16\ldots^{20}$
$= 3442.6\ldots$
Lake will be covered since $P(20) > 3400$.
Method 2: Find when $P(t) = 3400$.
$3400 = 150 \times 1.169\ldots^t$
$t = \frac{\log\left(\frac{3400}{150}\right)}{\log 1.169\ldots}$
$= 19.92\ldots$
Lake will be covered since 3400 plants are there before 20 days.

TEST 17 CHAPTER 6

1. $g(x) = 3^x$ is exponential since it has a variable in the exponent.

2. $-7^4 = \underline{\underline{-2401}}$

3. $5 \cdot 8 + 2 \cdot 4^3$
$= 40 + 128$
$= \underline{\underline{168}}$

4. a. $x = \log_b y$ if and only if $y = b^x$.
b. $\frac{x^a}{x^b} = x^{a-b}$
c. $\log xy = \log x + \log y$
d. $x^{\frac{1}{n}} = \sqrt[n]{x}$

5. $(3r^5 m^7)^4 (64r^0 m^{12})^{\frac{2}{3}}$
$= 81r^{20}m^{28} \cdot 16m^8$
$= \underline{\underline{1296r^{20}m^{36}}}$

6. $\sqrt[9]{32586.4} = \underline{\underline{3.172842\ldots}}$

7. $\frac{3.7495 \times 10^{-21}}{7.24 \times 10^{17}}$
$= 5.178867\ldots \times 10^{-39}$
$\approx \underline{\underline{5.18 \times 10^{-39}}}$

8. $f(x) = 7^x$
$x = 7^{f^{-1}(x)}$
$\underline{\underline{f^{-1}(x) = \log_7 x}}$

9. $\log_{25} x = \frac{3}{2}$
$25^{\frac{3}{2}} = x$
$\underline{\underline{125}} = x$

10. Let $x = \log_6 37.9$
$\therefore 6^x = 37.9$
$x = \frac{\log 37.9}{\log 6}$
$x = \underline{\underline{2.0287\ldots}}$

11. $\log_x 64 = \frac{2}{3}$
$x^{\frac{2}{3}} = 64$
$x = 64^{\frac{3}{2}}$
$x = \underline{\underline{512}}$

12. $3 \log_5 12 - \log_5 36$
$= \log_5 \frac{12^3}{36}$
$= \underline{\underline{\log_5 48}}$

13. 8.75×10^{29}
Characteristic $= \underline{\underline{29}}$
Mantissa $= \underline{\underline{8.75}}$

14.

x	$h(x)$
2	24
5 (+3)	36 (× 1.5)
8 (+3)	54 (× 1.5)
11 (+3)	81 (× 1.5)

$\therefore h(11) = \underline{\underline{81}}$

15. Example:

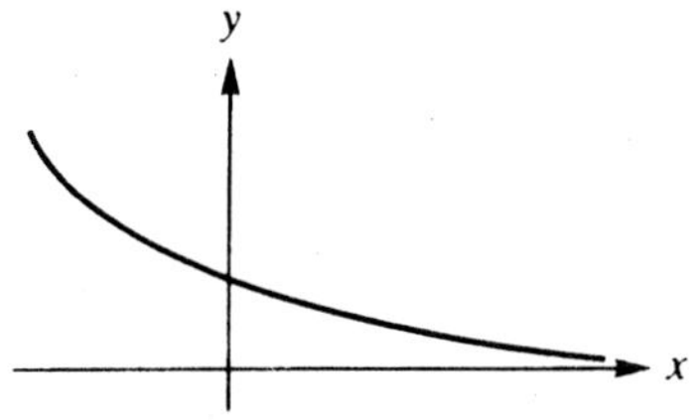

16. *Aspirin Problem*
Let A = no. of units of aspirin.
Let t = no. of minutes.
$A = C \times b^t$
(0, 100), (40, 73)
$C = 100$
$73 = 100 \times b^{40}$
$b = \left(\frac{73}{100}\right)^{\frac{1}{40}}$
$= 0.992163\ldots$
$\therefore \underline{\underline{A = 100 \times (0.992163\ldots)^t}}$
If $t = 95$,
$A = 100 \times 0.992\ldots^{95}$
$= 47.357\ldots$
About 47 units

TEST 18 CHAPTER 7 Sections 7-1 to 7-5

1. $12a^3b^5 + 18ac = \underline{\underline{6a(2a^2b^5 + 3c)}}$

2. $9x^2 - 36 = 9(x^2 - 4)$
$= \underline{\underline{9(x + 2)(x - 2)}}$

3. $x^2 + 14xy - 15y^2 = \underline{\underline{(x + 15y)(x - y)}}$

4. $4x^2 + 2x - 30 = 2(2x^2 + x - 15)$
$= \underline{\underline{2(2x - 5)(x + 3)}}$

5. $4z^2 + 25z + 6 = \underline{\underline{(4z + 1)(z + 6)}}$

6. $9x^2 - 48x + 64 = \underline{\underline{(3x - 8)^2}}$

7. $x^3 + c^3 = \underline{\underline{(x + c)(x^2 - xc + c^2)}}$

8. $8p^3 - f^3 = \underline{\underline{(2p - f)(4p^2 + 2pf + f^2)}}$

9. $(x + a)y + (x + a)(3p - 4)$
$= \underline{\underline{(x + a)(y + 3p - 4)}}$

10. $36x^2 + 72x + 25$
$b^2 - 4ac = 72^2 - 4(36)(25)$
$= 1584$

$\sqrt{1584} = 39.799\ldots$, which is not a perfect square.
Therefore, the polynomial is prime, Q.E.D.

11. $(3x - 7)(4x + 8) = \underline{\underline{12x^2 - 4x - 56}}$

12. $(x^2 - 7x + 3)(x - 4) = \underline{\underline{x^3 - 11x^2 + 31x - 12}}$

13. $(5x + 11)^2 = \underline{\underline{25x^2 + 110x + 121}}$

14. $(x - 3)(x + 5)(x - 8)$
$= (x^2 + 2x - 15)(x - 8)$
$= \underline{\underline{x^3 - 6x^2 - 31x + 120}}$

15.
$$\begin{array}{r}
4x^2 + 5x + 17 \\
x - 2\,\overline{)\,4x^3 - 3x^2 + 7x - 19} \\
\underline{4x^3 - 8x^2} \qquad\qquad \\
5x^2 + 7x \qquad \\
\underline{5x^2 - 10x} \qquad \\
17x - 19 \\
\underline{17x - 34} \\
15
\end{array}$$

$\therefore \dfrac{4x^3 - 3x^2 + 7x - 19}{x - 2} = \underline{\underline{4x^2 + 5x + 17 + \dfrac{15}{x - 2}}}$

16.
$$\begin{array}{r}
x^2 - 3x + 9 \\
x + 3\,\overline{)\,x^3 + 0x^2 + 0x - 11} \\
\underline{x^3 + 3x^2} \qquad\qquad \\
-3x^2 + 0x \qquad \\
\underline{-3x^2 - 9x} \qquad \\
9x - 11 \\
\underline{9x + 27} \\
-38
\end{array}$$

$\therefore \dfrac{x^3 - 11}{x + 3} = \underline{\underline{x^2 - 3x + 9 - \dfrac{38}{x + 3}}}$

17. $f(x) = \dfrac{1}{x - 3}$

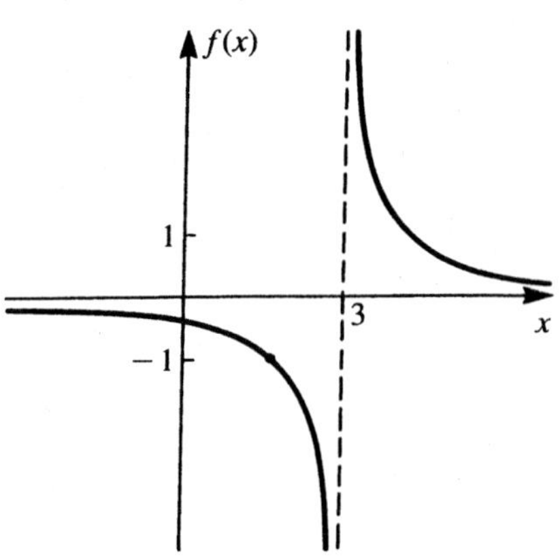

18. $g(x) = \dfrac{(x + 2)(x - 3)}{x + 2}$
$g(x) = x - 3$, provided $x \neq -2$.

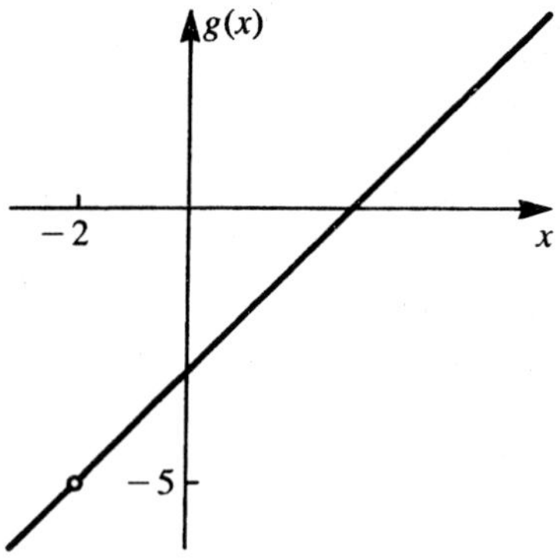

19. e.g.

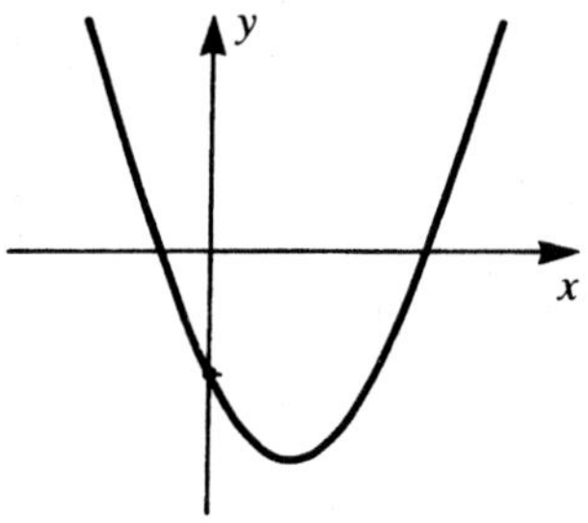

TEST 19 CHAPTER 7 Sections 7-6 to 7-9

1. $\dfrac{x^2 - 64}{x^2 - 16} \div \dfrac{x - 8}{x - 4}$
$= \dfrac{(x + 8)(x - 8)}{(x + 4)(x - 4)} \cdot \dfrac{x - 4}{x - 8}$
$= \underline{\underline{\dfrac{x + 8}{x + 4}}}$

2. $\dfrac{x - 7}{7} - \dfrac{x - 4}{4}$
$= \dfrac{4x - 28 - 7x + 28}{28}$
$= \underline{\underline{-\dfrac{3x}{28}}}$

3. $\dfrac{x}{x + y} + \dfrac{y}{x - y}$
$= \dfrac{x^2 - xy + xy + y^2}{(x + y)(x - y)}$
$= \underline{\underline{\dfrac{x^2 + y^2}{x^2 - y^2}}}$

4. $\dfrac{5x + 17}{x^2 + 8x + 7} - \dfrac{3}{x + 7}$
$= \dfrac{5x + 17 - 3x - 3}{(x + 7)(x + 1)}$
$= \dfrac{2x + 14}{(x + 7)(x + 1)}$
$= \underline{\underline{\dfrac{2}{x + 1}}}$

5. $\dfrac{20 + \dfrac{10}{x}}{20 - \dfrac{5}{x^2}} \cdot \dfrac{x^2}{x^2}$
$= \dfrac{20x^2 + 10x}{20x^2 - 5} = \dfrac{10x(2x + 1)}{5(2x + 1)(2x - 1)} = \underline{\underline{\dfrac{2x}{2x - 1}}}$

6. $P(x) = x^3 - 5x^2 - 2x + 24$

$P(-2) = -8 - 20 + 4 + 24 = 0$

$\therefore (x + 2)$ is a factor.

(*Note:* Synthetic substitution may be used, as shown here, rather than long division. See Section 10-5.)

$$\begin{array}{r|rrrr} -2 & 1 & -5 & -2 & 24 \\ & & -2 & 14 & -24 \\ \hline & 1 & -7 & 12 & 0 \end{array}$$

$\therefore P(x) = (x + 2)(x^2 - 7x + 12)$

$= \underline{\underline{(x + 2)(x - 3)(x - 4)}}$

7. a. $P(x) = 3x^3 + 13x^2 - 4x - 4$

$P\left(\frac{2}{3}\right) = 3 \cdot \frac{8}{27} + 13 \cdot \frac{4}{9} - 4 \cdot \frac{2}{3} - 4$

$= \frac{8}{9} + \frac{52}{9} - \frac{8}{3} - 4$

$= 0$, Q.E.D.

b. Factor is $\underline{\underline{(3x - 2)}}$.

8. $x^7 - y^7 = \underline{\underline{(x - y)(x^6 + x^5y + x^4y^2 + x^3y^3 + x^2y^4 + xy^5 + y^6)}}$

9. $a^3 + b^3 = \underline{\underline{(a + b)(a^2 - ab + b^2)}}$

10. $f(x) = \dfrac{5x - 10}{x^2 + x - 6} = \dfrac{5(x - 2)}{(x + 3)(x - 2)} = \dfrac{5}{x + 3}$,

$x \neq 2, -3$

x	$f(x)$
−8	−1
−7	−5/4
−6	−5/3
−5	−5/2
−4	−5
−3	no value
−2	5
−1	5/2
0	5/3
1	5/4
2	(1) removable disc.
3	5/6
4	5/7

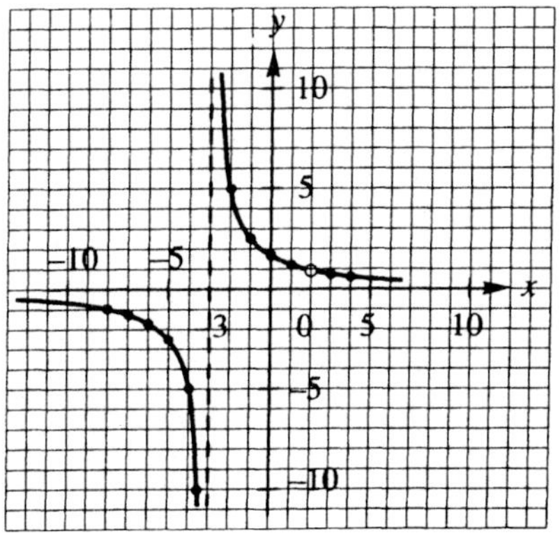

11. $f(x) = \dfrac{x^2 - 3x - 4}{x + 1} = \dfrac{(x + 1)(x - 4)}{x + 1}$

$= x - 4, x \neq -1$

Linear function,

$m = 1, b = -4$,

but removable discontinuity at

$x = -1, y = -5$.

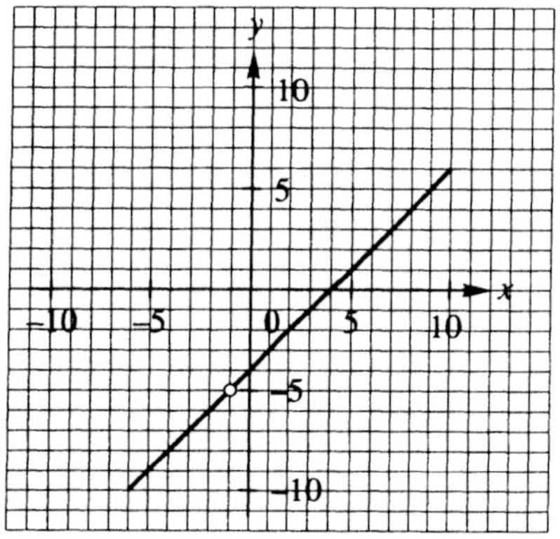

12. Let x = no. of ounces.

Let y = no. of cents.

(5.75, 89), (4, 75)

$m = \dfrac{89 - 75}{5.75 - 4} = 8$

$y - 75 = 8(x - 4)$

$\underline{\underline{y = 8x + 43}}$

$y = 147$

$147 = 8x + 43$

$13 = x$

$\underline{\underline{13 \text{ oz.}}}$

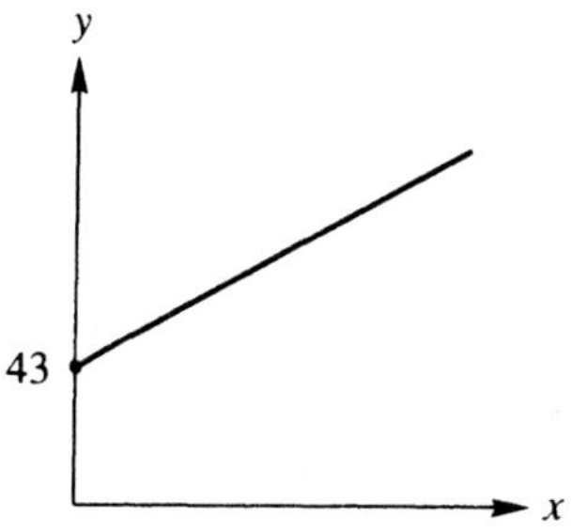

TEST 20 CHAPTER 7 Sections 7-10 and 7-11

1. $y = kx^5$

2. Example:

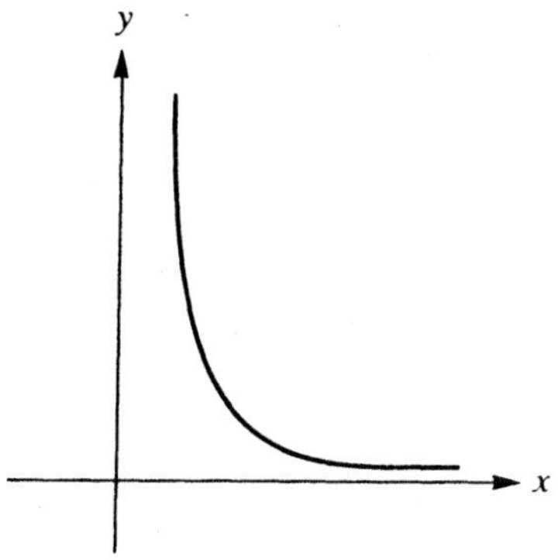

3. $p = \dfrac{k}{r}$, (54, 37)

$37 = \dfrac{k}{54}$

$1998 = k$

$\therefore$ equation is $\underline{\underline{p = \dfrac{1998}{r}}}$

4. $f(2) = 5$
$f(6) = 45$

Multiplying x by 3 multiplies $f(x)$ by 9, which is 3^2. So the function is a direct square variation.

$f(x) = kx^2$
$5 = k(2^2)$
$k = 1.25$
$\therefore f(x) = 1.25x^2$

5. a. Let h = no. of inches.
Let w = no. of pounds.

$w = kh^3$, (60, 99)
$99 = k \cdot 60^3$
$k = \frac{99}{60^3} = 0.0004583\ldots$
$\therefore$ equation is $w = 0.0004583\ldots h^3$

b. $w = (0.0004583\ldots)(76^3)$
$= 201.197\ldots$
About 201 lb. Wow!

c. $80 = 0.0004583\ldots h^3$
$h = \left(\frac{80}{0.0004583\ldots}\right)^{\frac{1}{3}}$
$h = 55.88\ldots$
About 56 inches

6. $1 + \frac{x}{x-3} = \frac{5}{x-3}, x \neq 3$
$x - 3 + x = 5$
$2x = 8$
$x = 4$
$S = \{4\}$

7. $\frac{5}{x+1} + \frac{x}{x-2} = \frac{8x-7}{x^2-x-2}, x \neq -1, 2$
$5x - 10 + x^2 + x = 8x - 7$
$x^2 - 2x - 3 = 0$
$(x-3)(x+1) = 0$
$x = 3$ or ~~$x = -1$~~ → Extraneous
$S = \{3\}$

8. $f(x) = \frac{12(x-1)(x-2)}{(x-1)(x+1)(x-4)}, x \neq 1, -1, 4$

x	$f(x)$
−6	−1.92
−5	$-2\frac{1}{3}$
−4	−3
−3	−4.28
−2	−8
−1	none
0	6
1	(2) Rem. discont.
2	0
3	−3
4	none
5	6
6	3.43
7	2.5
8	2

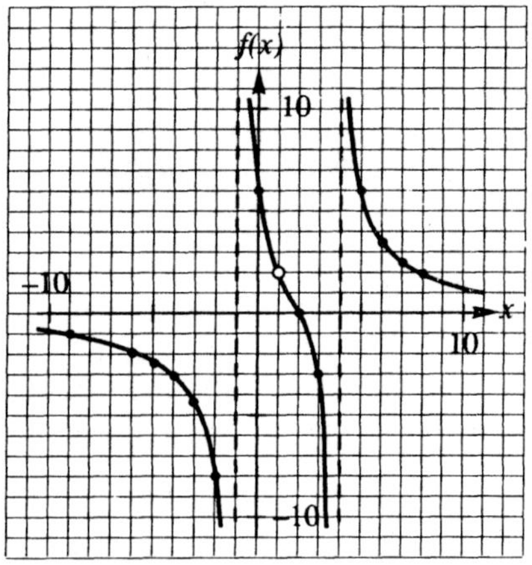

TEST 21 CHAPTER 7

1. $\frac{(x+3)(x-5)}{(x-4)(x-1)} \div \frac{(x+3)(5-x)}{(x+4)(x+7)}$
$= \frac{(x+3)(x-5)(x+4)(x+7)}{(x-4)(x-1)(x+3)(5-x)}$
$= -\frac{(x+4)(x+7)}{(x-4)(x-1)}$ or $-\frac{x^2+11x+28}{x^2-5x+4}$

2. $\frac{x}{x-2} - \frac{3}{x+4}$
$= \frac{x^2+4x-3x+6}{(x-2)(x+4)}$ $= \frac{x^2+x+6}{x^2+2x-8}$

3. $\frac{x^5+y^5}{x+y} = \frac{(x+y)(x^4-x^3y+x^2y^2-xy^3+y^4)}{x+y}$
$= x^4 - x^3y + x^2y^2 - xy^3 + y^4$

4. $\frac{x + \frac{2}{x+3}}{2 + \frac{x}{x+3}} \cdot \frac{x+3}{x+3}$
$= \frac{x^2+3x+2}{2x+6+x}$
$= \frac{(x+1)(x+2)}{3(x+2)}$
$= \frac{x+1}{3}$

5. $(3x-7)(x^2-4x+5)$
$= 3x^3 - 19x^2 + 43x - 35$

6. $x^3 + 125$
$= (x+5)(x^2-5x+25)$

7. $64x^2 - 16 = 16(4x^2-1)$
$= 16(2x+1)(2x-1)$

8. $P(x) = x^3 - 5x^2 - 8x + 48$
a. $P(-3) = -27 - 45 + 24 + 48 = 0$, Q.E.D.
b. $(x+3)$ is a factor. By synthetic substitution:

−3	1	−5	−8	48
		−3	24	−48
	1	−8	16	0

$P(x) = (x+3)(x^2-8x+16)$
$P(x) = (x+3)(x-4)^2$

9.
$$\begin{array}{r} 3x^2 + 4x - 5 \\ x - 6\,\overline{)3x^3 - 14x^2 - 29x + 5} \\ \underline{3x^3 - 18x^2} \qquad\qquad \\ 4x^2 - 29x \qquad \\ \underline{4x^2 - 24x} \qquad \\ -5x + 5 \\ \underline{-5x + 30} \\ -25 \end{array}$$

$$\therefore \frac{3x^3 - 14x^2 - 29x + 5}{x - 6} = \underline{\underline{3x^2 + 4x - 5 - \frac{25}{x - 6}}}$$

10. $x + \frac{1}{x - 1} = \frac{x}{x - 1}, x \neq 1$

$x^2 - x + 1 = x$

$x^2 - 2x + 1 = 0$

$(x - 1)^2 = 0$

~~$x = 1$~~ ← Extraneous

$S = \emptyset$

11. a. $y = \frac{k}{x^2}$

b. $y = kx^3$

12. $y = kx$, (9, 63)

$63 = k \cdot 9$

$7 = k$

$\therefore$ *equation is* $\underline{\underline{y = 7x}}$

13. $y = a \cdot b^x$

(0, 234), (6, 5810)

$a = 234$

$5810 = 234 \cdot b^x$

$\left(\frac{5810}{234}\right)^{\frac{1}{6}} = b$

$b = 1.708 \ldots$

$y = 234 \cdot (1.708\ldots)^x$

14. $f(x) = \frac{12(x - 1)}{(x - 1)(x + 2)} = \frac{12}{x + 2}, x \neq 1, -2$

x	$f(x)$
−8	−2
−7	−2.4
−6	−3
−5	−4
−4	−6
−3	−12
−2	none
−1	12
−0	6
1	(4) rem. discont.
2	3
3	2.4
4	2

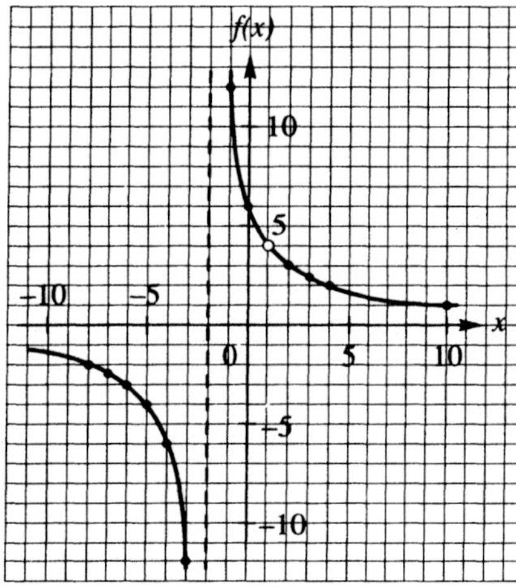

TEST 22 CHAPTER 8

1. $\sqrt{18} + 2\sqrt{50} - \sqrt{98}$

$= 3\sqrt{2} + 10\sqrt{2} - 7\sqrt{2}$

$= \underline{\underline{6\sqrt{2}}}$

2. $\frac{21}{\sqrt{12} + \sqrt{5}} \cdot \frac{\sqrt{12} - \sqrt{5}}{\sqrt{12} - \sqrt{5}}$

$= \frac{21(\sqrt{12} - \sqrt{5})}{12 - 5}$

$= 3(2\sqrt{3} - \sqrt{5})$

$= \underline{\underline{6\sqrt{3} - 3\sqrt{5}}}$

3. $\frac{1 - \frac{1}{\sqrt{3}}}{1 + \frac{1}{\sqrt{3}}} \cdot \frac{\sqrt{3}}{\sqrt{3}}$

$= \frac{\sqrt{3} - 1}{\sqrt{3} + 1} \cdot \frac{\sqrt{3} - 1}{\sqrt{3} - 1}$

$= \frac{3 - 2\sqrt{3} + 1}{3 - 1}$

$= \underline{\underline{2 - \sqrt{3}}}$

4. $\sqrt{x + 3} + \sqrt{x - 2} = 5$

$\sqrt{x + 3} = 5 - \sqrt{x - 2}$

$x + 3 = 25 - 10\sqrt{x - 2} + x - 2$

$10\sqrt{x - 2} = 20$

$x - 2 = 4$

$x = 6$

Check: $\sqrt{6 + 3} + \sqrt{6 - 2} = 5$ ✓

$S = \{6\}$

5. *Labor Problem*

Let t = no. of days.

Let w = no. of workers.

$t = kw^{-0.7}$, (100, 250)

$250 = k(100^{-0.7})$

$k = 250(100^{0.7})$

$= 6279.71\ldots$

$\therefore$ equation is $\underline{\underline{t = 6279.71\ldots w^{-0.7}}}$

If $t = 180$, then

$180 = 6279.71\ldots w^{-0.7}$

$w = \left(\frac{6279.71\ldots}{180}\right)^{\frac{1}{0.7}}$

$w = 159.88\ldots$

<u>About 160 workers</u>

6. *Bridge Problem*

Let c = no. of dollars cost.

Let d = no. of feet deep.

Let L = no. of feet long.

$c = kLd^2$

$700{,}000 = k \cdot 200 \cdot 8^2$

$54.6875 = k$

$\therefore$ *equation is* $c = 54.6875\ Ld^2$

$c = (54.6875)(600)(5^2)$

$= 820{,}312.5$

About \$820,000

7. $f(x) = x - \sqrt{x - 2}$

x	$f(x)$
11	8
6	4
3	2
2.5	1.792...
2	2
0	no real value

8.

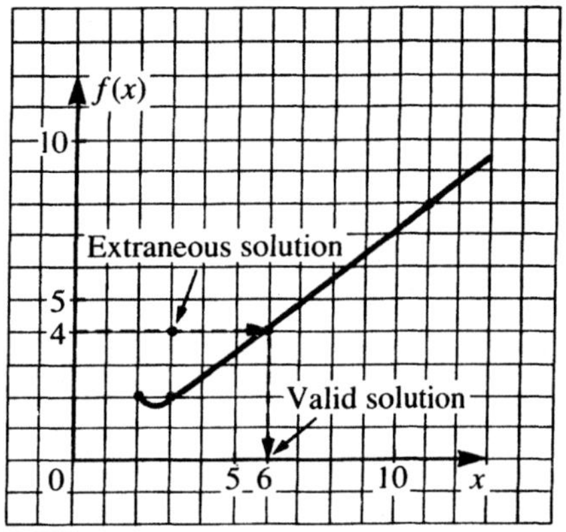

9. Domain = $\{x: x \geq 2\}$

10. $f(x) = 0$:

$0 = x - \sqrt{x - 2}$

$\sqrt{x - 2} = x$

$x - 2 = x^2$

$0 = x^2 - x + 2 = 0$

$b^2 - 4ac = 1 - 4 \cdot 1 \cdot 2 = -7$

$\therefore$ There are no real values of x.

11. $f(x) = 4$:

$4 = x - \sqrt{x - 2}$

$\sqrt{x - 2} = x - 4$

$x - 2 = x^2 - 8x + 16$

$0 = x^2 - 9x + 18$

$(x - 3)(x - 6) = 0$

$x = 3$ or $x = 6$

Checks:

$4 \stackrel{?}{=} 3 - \sqrt{1}$ $\quad$ $4 \stackrel{?}{=} 6 - \sqrt{4}$

$4 \neq 2$ $\quad$ $4 = 4\ \checkmark$

$x = 3$ is extraneous. $x = 6$ is valid.

See graph.

TEST 23 CHAPTERS 6 through 8

1. a. Example:

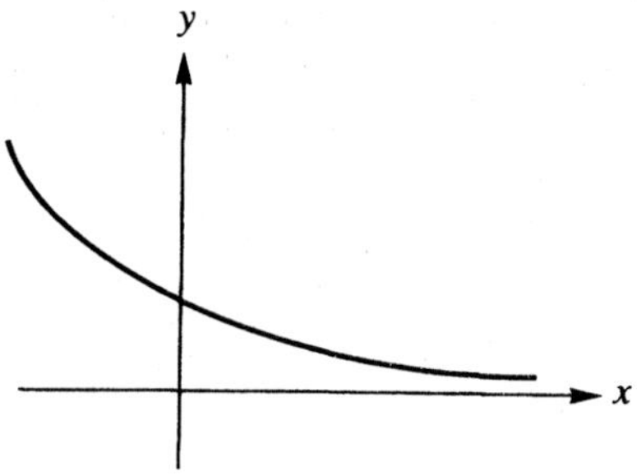

b. Example:

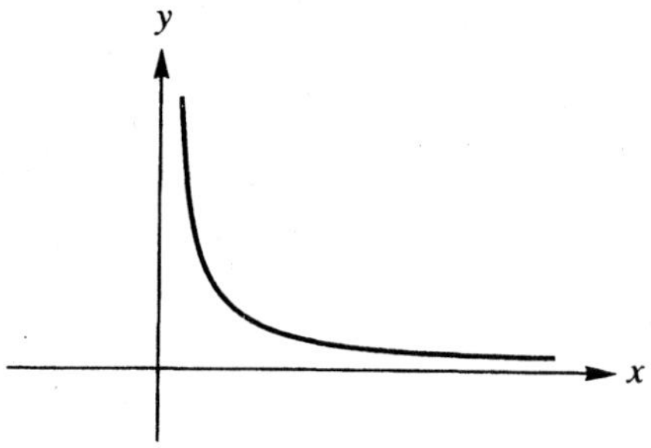

c. Example:

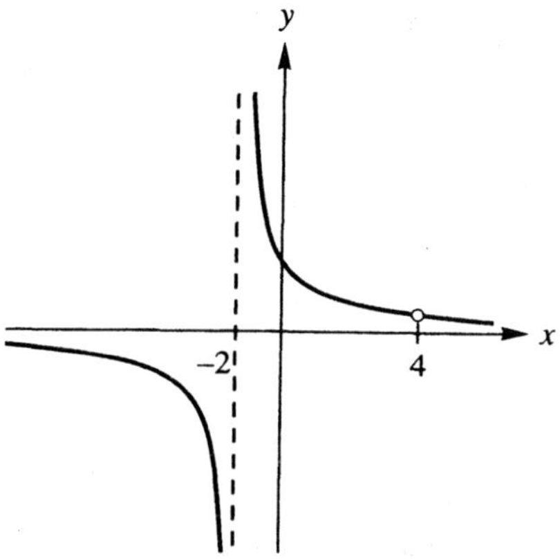

d. Example:

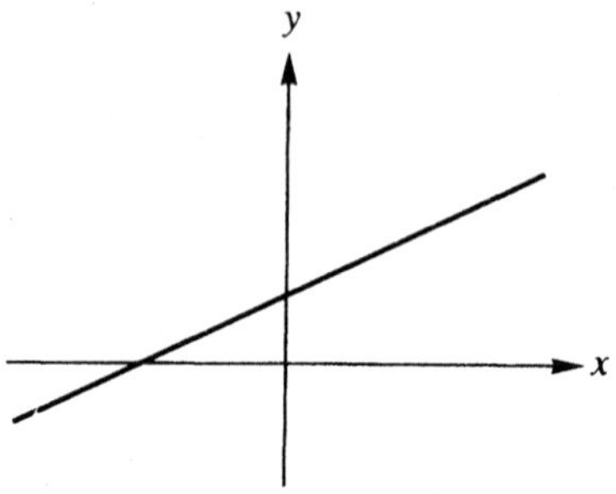

e. Example:

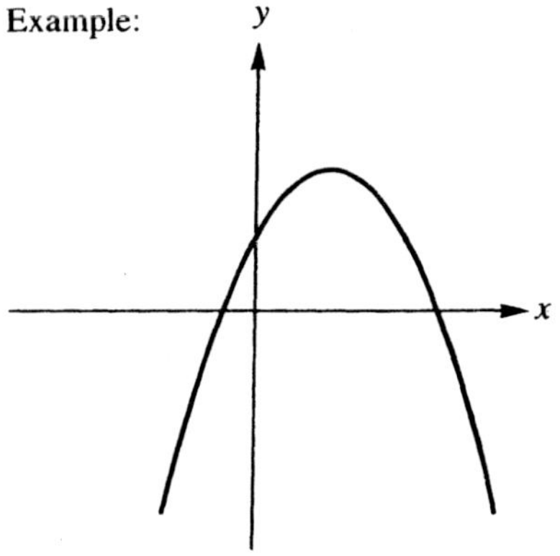

2. a. $y = kx^3$ b. $y = \dfrac{k}{x^2}$

c. $y = a \cdot b^x$ or $y = c \times 10^{kx}$

d. $y = mx + b$

3. a. $(\sqrt{8} + \sqrt{3})(\sqrt{15} + \sqrt{2})$

$= \underline{\underline{2\sqrt{30} + 4 + 3\sqrt{5} + \sqrt{6}}}$

b. $(\sqrt{3} - \sqrt{5})^2$

$= 3 - 2\sqrt{15} + 5$

$= \underline{\underline{8 - 2\sqrt{15}}}$

c. $(\sqrt{y} + 6)(\sqrt{y} - 6)$

$= \underline{\underline{y - 36}}$

d. $(3x + 11)(2x - 7)$

$= \underline{\underline{6x^2 + x - 77}}$

e. $(x + 3)(x^2 - 4x - 5)$

$= \underline{\underline{x^3 - x^2 - 17x - 15}}$

4. a. $a^5 + b^5$

$= \underline{\underline{(a + b)(a^4 - a^3b + a^2b^2 - ab^3 + b^4)}}$

b. $5x^2 + 27x + 36$

$= \underline{\underline{(5x + 12)(x + 3)}}$

c. $x^3 + 2x^2 - 11x - 12$

$= (x + 1)(x^2 + x - 12)$

$= \underline{\underline{(x + 1)(x - 3)(x + 4)}}$

d. $64p^2 - 36$

$= 4(16p^2 - 9) = \underline{\underline{4(4p + 3)(4p - 3)}}$

5. a. $\dfrac{1}{x^2 + 4x + 3} + \dfrac{1}{x^2 - 1}$

$= \dfrac{1}{(x + 1)(x + 3)} \cdot \dfrac{x - 1}{x - 1}$

$+ \dfrac{1}{(x + 1)(x - 1)} \cdot \dfrac{x + 3}{x + 3}$

$= \dfrac{x - 1 + x + 3}{(x + 1)(x + 3)(x - 1)}$

$= \dfrac{2(x + 1)}{(x + 1)(x + 3)(x - 1)}$

$= \underline{\underline{\dfrac{2}{x^2 + 2x - 3}}}$

b. $\dfrac{9 - x^2}{3x + 9} \div \dfrac{x^2 - 5x + 6}{x^2 + 2x + 1}$

$= \dfrac{(3 + x)(3 - x)}{3(x + 3)} \cdot \dfrac{(x + 1)^2}{(x - 2)(x - 3)}$

$= \underline{\underline{-\dfrac{(x + 1)^2}{3(x - 2)}}}$ or $\underline{\underline{-\dfrac{x^2 + 2x + 1}{3x - 6}}}$

6. a. $5x^{-\frac{2}{5}} \cdot 7x^{\frac{3}{4}} = \underline{\underline{35x^{\frac{7}{20}}}}$

b. $\sqrt{125} \div \sqrt[3]{25}$

$= 5^{\frac{3}{2}} \div 5^{\frac{2}{3}}$

$= \underline{\underline{5^{\frac{5}{6}}}}$ or $\underline{\underline{\sqrt[6]{3125}}}$

c. $\dfrac{12}{\sqrt{19} + 4} \cdot \dfrac{\sqrt{19} - 4}{\sqrt{19} - 4}$

$= \dfrac{12(\sqrt{19} - 4)}{19 - 16}$

$= \underline{\underline{4\sqrt{19} - 16}}$

d. $\dfrac{24}{\sqrt[3]{2}} \cdot \dfrac{\sqrt[3]{4}}{\sqrt[3]{4}}$

$= \dfrac{24\sqrt[3]{4}}{2} = \underline{\underline{12\sqrt[3]{4}}}$

e. $\sqrt[6]{36} = (6^2)^{\frac{1}{6}} = 6^{\frac{1}{3}} = \underline{\underline{\sqrt[3]{6}}}$

f. $7\sqrt{45} + \dfrac{10}{\sqrt{5}} \cdot \dfrac{\sqrt{5}}{\sqrt{5}}$

$= 21\sqrt{5} + 2\sqrt{5} = \underline{\underline{23\sqrt{5}}}$

g. $\log_9 27 = x$

$9^x = 27$

$3^{2x} = 3^3$

$x = \underline{\underline{\dfrac{3}{2}}}$

h. $\log_7 30 + 2 \log_7 5$

$= \log_7(30 \times 5^2)$

$= \underline{\underline{\log_7 750}}$

i. $\left(\dfrac{5}{x - 2} + 1\right) \div \left(\dfrac{5}{x + 3} - 1\right)$

$= \dfrac{5 + x - 2}{x - 2} \cdot \dfrac{x + 3}{5 - x - 3}$

$= \dfrac{(x + 3)(x + 3)}{(x - 2)(2 - x)}$

$= -\dfrac{(x + 3)^2}{(x - 2)^2}$ (*Note:* No canceling is possible!)

7. a. $3^{2x} = 457$

$2x \cdot \log 3 = \log 457$

$x = \dfrac{\log 457}{2 \cdot \log 3}$

$x = 2.78746\ldots$

$S = \{2.78746\ldots\}$

b. $2 + \sqrt{x - 2} = x$

$\sqrt{x - 2} = x - 2$

$x - 2 = x^2 - 4x + 4$

$0 = x^2 - 5x + 6$

$0 = (x - 2)(x - 3)$

$x = 2$ or $x = 3$

check: $2 + \sqrt{2 - 2} = 2$ ✓

check: $2 + \sqrt{3 - 2} = 3$ ✓

$S = \{2, 3\}$

c. $x + \dfrac{1}{x - 2} = \dfrac{x - 3}{2 - x}, x \neq 2$

$x^2 - 2x + 1 = -x + 3$

$x^2 - x - 2 = 0$

$(x - 2)(x + 1) = 0$

~~$x = 2$~~ ext. or $x = -1$

$S = \{-1\}$

8. $f(x) = \dfrac{x^3 + 2x^2 - 11x - 12}{x^2 - x - 6}$

(*Note:* Numerator is the same as in Problem 4c.)

$$f(x) = \frac{(x+1)(x-3)(x+4)}{(x-3)(x+2)}$$

$$f(x) = \frac{(x+1)(x+4)}{x+2},\ x \neq 3$$

If $x = 3$, $\frac{(3+1)(3+4)}{3+2} = \frac{28}{5}$

Removable discontinuity is at $\left(3, 5\frac{3}{5}\right)$

9. *Bacteria Problem*

Let t = no. of days.

Let B = no. of millions of bacteria.

$B = a \cdot b^t$

(0, 56), (2, 204)

$a = 56$

$204 = 56 \cdot b^2$

$b = \left(\frac{204}{56}\right)^{\frac{1}{2}}$

$b = 1.90862...$

$\therefore$ equation is $B = 56 \times 1.908...^t$

If $t = 7$, $B = 56 \times 1.908...^7$

$= 5166.96...$

About 5167 million

10. *Braking Distance Problem*

a. Let s = no. of mph.

Let f = coefficient of friction.

Let d = no. of feet.

$d = \frac{ks^2}{f}$

$45 = \frac{k \cdot 30^2}{0.8}$

$k = 0.04$

$\therefore$ equation is $d = \frac{0.04\, s^2}{f}$

b. $290 = \frac{0.04s^2}{0.3}$

$s = \left(\frac{(290)(0.3)}{0.04}\right)^{\frac{1}{2}}$

$s = 46.63...$

About 47 mph.

11. *Christmas Tree Problem*

Let h = no. of ft high.

Let p = no. of dollars price.

$p = kh^{2.3}$, (6, 48)

$48 = k(6^{2.3})$

$k = 0.77892...$

$\therefore$ equation is $p = 0.77892...\ h^{2.3}$

If $h = 60$, then

$p = (0.77892...)(60^{2.3})$

$p = 9577.259...$

About \$9600 (High, but that's about what they would charge!)

TEST 24 CHAPTER 9 Sections 9-1 to 9-5

1. $x^2 + y^2 + 4x - 10y + 20 < 0$

$(x^2 + 4x + 4) + (y^2 - 10y + 25) < -20 + 4 + 25$

$(x+2)^2 + (y-5)^2 < 9$

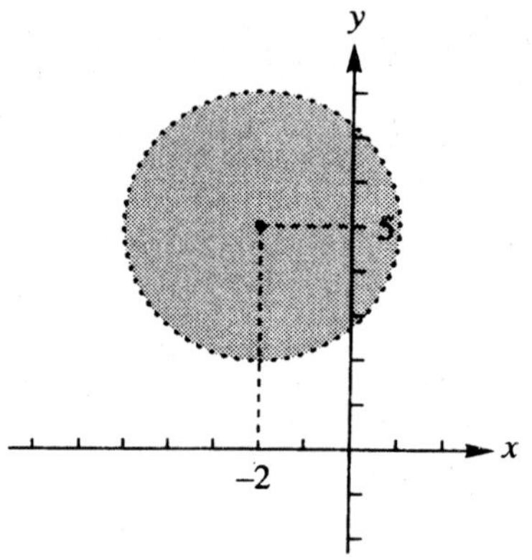

2. $4x^2 + y^2 - 8x + 6y - 3 = 0$

$4(x^2 - 2x + 1) + (y^2 + 6y + 9) = 3 + 4 + 9$

$4(x-1)^2 + (y+3)^2 = 16$

$\left(\frac{x-1}{2}\right)^2 + \left(\frac{y+3}{4}\right)^2 = 1$

$c^2 = 4^2 - 2^2 = 12$

$c = \sqrt{12} = 3.46...$

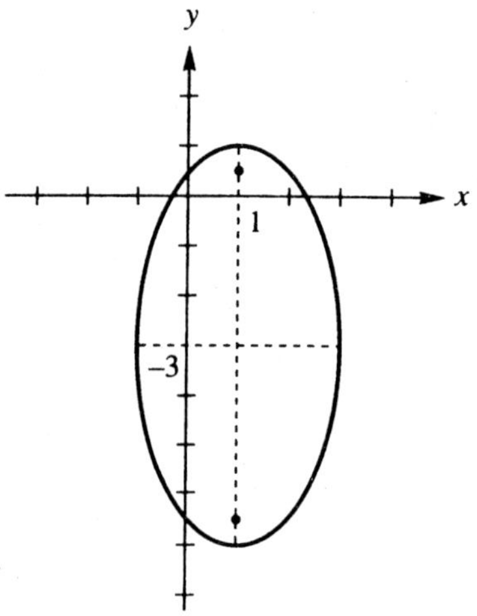

3. $y^2 + 2x - 6y - 7 = 0$

$2x - 7 = -y^2 + 6y$

$2x - 7 - 9 = -(y^2 - 6y + 9)$

$x - 8 = -\frac{1}{2}(y-3)^2$

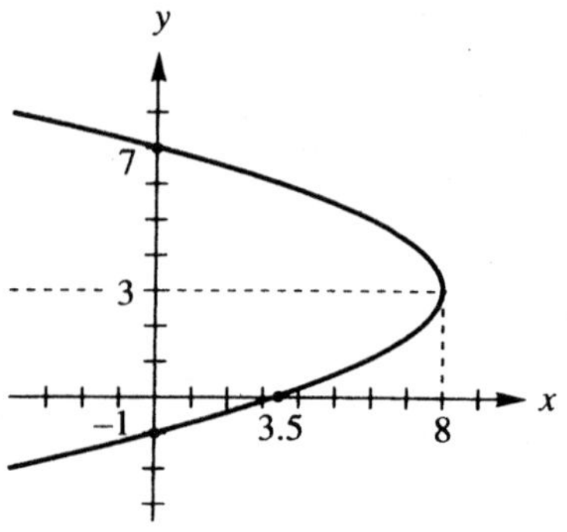

4. Equation is:

$$-\left(\frac{x-4}{2}\right)^2 + \left(\frac{y-5}{3}\right)^2 = 1$$

$$-9(x^2 - 8x + 16) + 4(y^2 - 10y + 25) = 36$$

$$\underline{\underline{9x^2 - 4y^2 - 72x + 40y + 80 = 0}}$$

5. $\frac{-9}{x-2} - x = 8, \quad x \neq 2$

$$-9 - x^2 + 2x = 8x - 16$$

$$0 = x^2 + 6x - 7$$

$$0 = (x + 7)(x - 1)$$

$$x = -7, x = 1$$

$$S = \{-7, 1\}$$

6. $\sqrt{x^2 + 35} - x = -7$

$$\sqrt{x^2 + 35} = x - 7$$

$$x^2 + 35 = x^2 - 14x + 49$$

$$14x = 14$$

~~$x = 1$~~ ← extraneous

check: $\sqrt{1 + 35} - 1 \neq -7$

$$S = \emptyset$$

7. $\frac{22}{\sqrt{18} + \sqrt{7}} \cdot \frac{\sqrt{18} - \sqrt{7}}{\sqrt{18} - \sqrt{7}}$

$$= \frac{22(3\sqrt{2} - \sqrt{7})}{18 - 7}$$

$$= \underline{\underline{6\sqrt{2} - 2\sqrt{7}}}$$

TEST 25 CHAPTER 9 Sections 9-6 to 9-8

1. a.

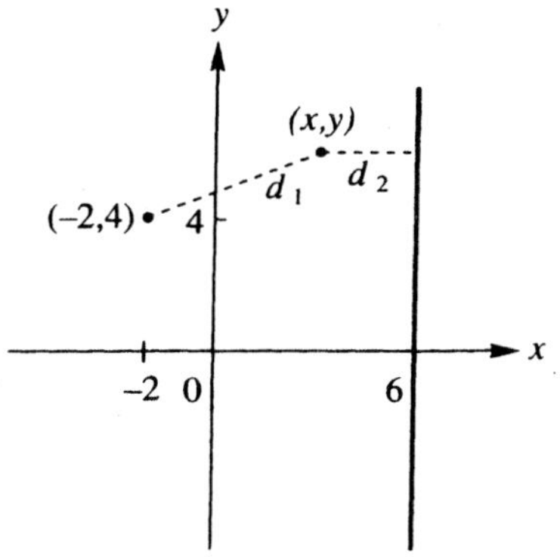

$$d_1 = 3d_2$$

$$\sqrt{(x+2)^2 + (y-4)^2} = 3|x - 6|$$

$$(x^2 + 4x + 4) + (y^2 - 8y + 16) = 9(x^2 - 12x + 36)$$

$$\underline{\underline{8x^2 - y^2 - 112x + 8y + 304 = 0}}$$

b. <u>Hyperbola.</u>

2. $4x^2 + 9y^2 + 7x - 2y - 100 = 0$

$B^2 - 4AC = 0 - 4(4)(9) = \underline{\underline{-144}}$

<u>Ellipse</u>

3. $4x^2 + 10xy + 9y^2 + 7x - 2y - 100 = 0$

$B^2 - 4AC = 100 - 4(4)(9) = \underline{\underline{-44}}$

<u>Ellipse</u>

4. $4x^2 + 12xy + 9y^2 + 7x - 2y - 100 = 0$

$B^2 - 4AC = 144 - 4(4)(9) = \underline{\underline{0}}$

<u>Parabola</u>

5. $4x^2 + 14xy + 9y^2 + 7x - 2y - 100 = 0$

$B^2 - 4AC = 196 - 4(4)(9) = \underline{\underline{52}}$

<u>Hyperbola</u>

6. $4x^2 + 14xy + 7x - 2y - 100 = 0$

$B^2 - 4AC = 196 - 4(4)(0) = \underline{\underline{196}}$

<u>Hyperbola</u>

7. $7x^2 - 5y^2 + 20y = 3$ ①

$21x^2 + 5y^2 = 209$ ②

$3 \times$ ①: $21x^2 - 15y^2 + 60y = 9$

②: $21x^2 + 5y^2 = 209$

$$20y^2 - 60y = 200$$

$$y^2 - 3y - 10 = 0$$

$$(y - 5)(y + 2) = 0$$

$$y = 5 \quad \text{or} \quad y = -2$$

$y = 5$:	$y = -2$:
$21x^2 + 125 = 209$	$21x^2 + 20 = 209$
$x^2 = 4$	$x^2 = 9$
$x = \pm 2$	$x = \pm 3$

$$S = \{(2, 5), (-2, 5), (3, -2), (-3, -2)\}$$

8. $x^2 + y^2 - 4x + 2y = 20$ ①

$4x + 3y = 5$ ②

From ②, $y = \frac{1}{3}(5 - 4x)$ ③

Substitute in ①:

$$x^2 + \frac{1}{9}(5 - 4x)^2 - 4x + \frac{2}{3}(5 - 4x) = 20$$

$$9x^2 + 25 - 40x + 16x^2 - 36x + 30 - 24x = 180$$

$$25x^2 - 100x - 125 = 0$$

$$x^2 - 4x - 5 = 0$$

$$(x - 5)(x + 1) = 0$$

$$x = 5 \quad \text{or} \quad x = -1$$

$x = 5$	$x = -1$
$y = \frac{1}{3}(5 - 20)$	$y = \frac{1}{3}(5 + 4)$
$y = -5$	$y = 3$

$$\underline{\underline{S = \{(5, -5), (-1, 3)\}}}$$

①: $(x^2 - 4x + 4) + (y^2 + 2y + 1) = 20 + 4 + 1$

$(x - 2)^2 + (y + 1)^2 = 25$

Center: $(2, -1)$ Radius: 5

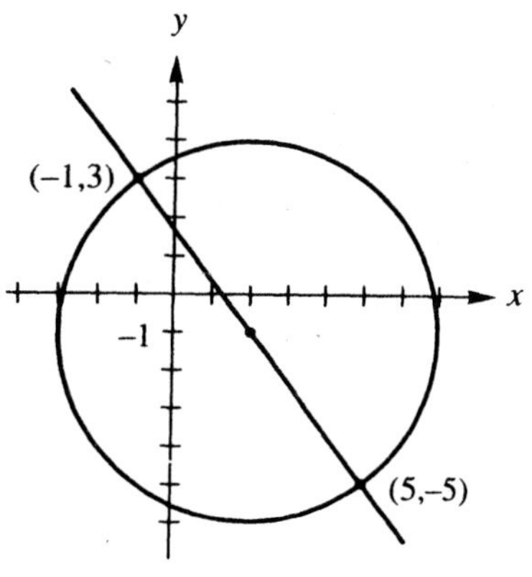

(*Note:* Problems 7 and 8 are Examples 2 and 4, respectively, in Section 9-8.)

TEST 26 CHAPTER 9

1. $x^2 + y^2 = 36$

Circle, since x^2 and y^2 have equal coefficients.

2. Center is at $(0, 0)$

3. a. $x^2 + y^2 > 36$

b. Example: $x^2 + y^2 = 49$

c. Example: $x^2 + x + y^2 = 36$

d. Example: $x^2 + 2y^2 = 36$

e. Example: $x^2 - y^2 = 36$

f. Example: $x^2 + xy + y^2 = 36$

4.

$$16x^2 - 5y^2 = 64 \quad ①$$
$$16x^2 + 25y^2 - 96x = 256 \quad ②$$
$$5 \times ①: 80x^2 - 25y^2 = 320$$
$$②: 16x^2 + 25y^2 - 96x = 256$$
$$96x^2 - 96x = 576$$
$$x^2 - x - 6 = 0$$
$$(x - 3)(x + 2) = 0$$
$$x = 3, x = -2$$
$$x = 3: 144 - 5y^2 = 64$$
$$-5y^2 = -80$$
$$y = \pm 4$$
$$x = -2: 64 - 5y^2 = 64$$
$$-5y^2 = 0$$
$$y = 0$$

Points are $(3, 4), (3, -4), (-2, 0)$

5. Center: $(-4, 2)$

$a = 7, b = 3$

$$\therefore \left(\frac{x + 4}{7}\right)^2 + \left(\frac{y - 2}{3}\right)^2 = 1$$

6. a. $\sqrt{(x - 3)^2 + (y - 4)^2} = 4\sqrt{(x + 5)^2 + (y - 2)^2}$

$x^2 - 6x + 9 + y^2 - 8y + 16 =$

$16(x^2 + 10x + 25 + y^2 - 4y + 4)$

$15x^2 + 15y^2 + 166x - 56y + 439 = 0$

b. Circle

7. $x = \frac{1}{4}y^2 - 2y - 1$

$x + 1 + 4 = \frac{1}{4}(y^2 - 8y + 16)$

$x + 5 = \frac{1}{4}(y - 4)^2$ Vertex: $(-5, 4)$

Opens in x-direction.

x-intercept $= -1$

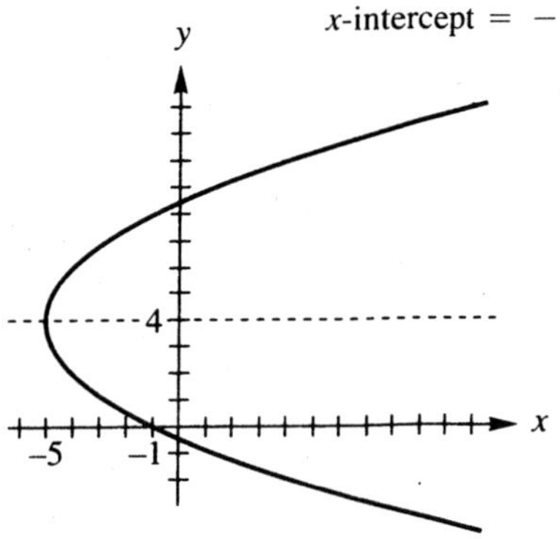

8. $x^2 - 9y^2 = 36$

$$\left(\frac{x}{6}\right)^2 - \left(\frac{y}{2}\right)^2 = 1$$

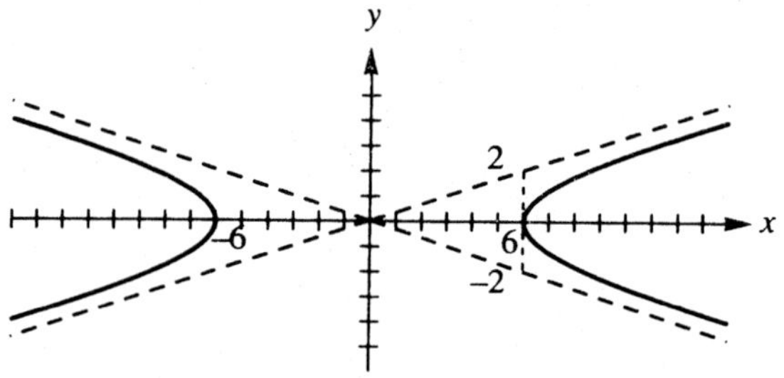

TEST 27 CHAPTER 10 Sections 10-1 to 10-3

1. $i^{59} = i^3 = -i$

2. $i^{-15} = \frac{1}{i^{15}} = \frac{1}{i^3} = i$

3. $i^{500} = 1$

4. $\sqrt{-18} \times \sqrt{-25}$

$= 3i\sqrt{2} \times 5i$

$= -15\sqrt{2}$

5. $\sqrt{(-5)(-35)}$

$= \sqrt{175}$

$= 5\sqrt{7}$

6. $\frac{\sqrt{98}}{\sqrt{-2}} = \frac{7\sqrt{2}}{i\sqrt{2}} = -7i$

7. $\sqrt{-12} + \frac{12}{\sqrt{-3}} \cdot \frac{i\sqrt{3}}{i\sqrt{3}}$

$= 2i\sqrt{3} + \frac{12i\sqrt{3}}{-3}$

$= 2i\sqrt{3} - 4i\sqrt{3}$

$= -2i\sqrt{3}$

8. $(5 - 2i) + (7 + 6i)$

$= 12 + 4i$

9. $(-8 + 2i) - (3 - 11i)$

$= -11 + 13i$

10. $(11 + 4i)(10 - 3i)$

$= 110 + 7i + 12$

$= 122 + 7i$

11. $(12 + i)(12 - i)$

$= 144 + 1$

$= 145$

12 $(12 + i)^2$

$= 144 + 24i - 1$

$= 143 + 24i$

13. $\frac{12 + 9i}{3 - 4i} \cdot \frac{3 + 4i}{3 + 4i}$

$= \frac{36 + 75i - 36}{9 + 16}$

$= 3i$

14. $x^2 - 6x + 58$

$= (x - 3 - 7i)(x - 3 + 7i)$

15. $x^2 + 10x + 29 = 0$

$x = \frac{-10 \pm \sqrt{100 - 4(1)(29)}}{2}$

$x = \frac{-10 \pm \sqrt{-16}}{2}$

$x = -5 \pm 2i$

$S = \{-5 + 2i, -5 - 2i\}$

16. $3x^2 - 6x + 4 = 0$

$x = \frac{6 \pm \sqrt{36 - 4(3)(4)}}{2(3)}$

$x = 1 \pm \frac{i\sqrt{3}}{3}$

$S = \left\{1 + \frac{i\sqrt{3}}{3}, 1 - \frac{i\sqrt{3}}{3}\right\}$

17. $16x^2 + 49 = 0$

$x^2 = -\frac{49}{16}$

$S = \left\{\frac{7}{4}i, -\frac{7}{4}i\right\}$

18. Check $x = -5 + 2i$
$(-5 + 2i)^2 + 10(-5 + 2i) + 29 \stackrel{?}{=} 0$
$25 - 20i - 4 - 50 + 20i + 29 \stackrel{?}{=} 0$
$0 = 0$

19. Solutions: $3, -5$
$\therefore \underline{\underline{x^2 + 2x - 15 = 0}}$

20. One solution is $6 + 4i$.
$\therefore$ other solution is $6 - 4i$.
Equation is $\underline{\underline{x^2 - 12x + 52 = 0}}$

21. $x^2 - y^2 + 3x - 5y + 13 = 0$
Hyperbola

TEST 28 CHAPTER 10 Sections 10-4 to 10-6

1. a. Example:

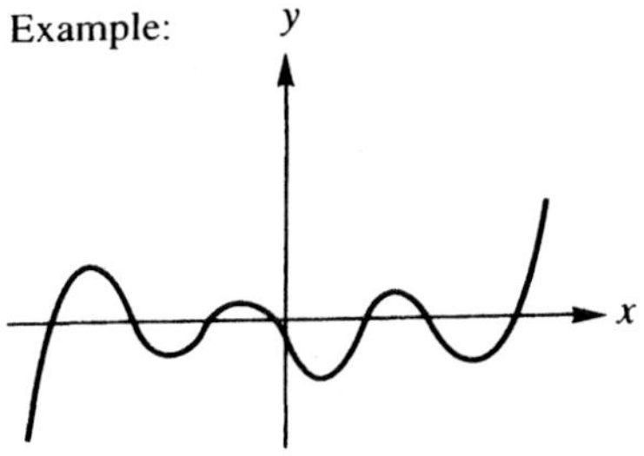

b. Example:

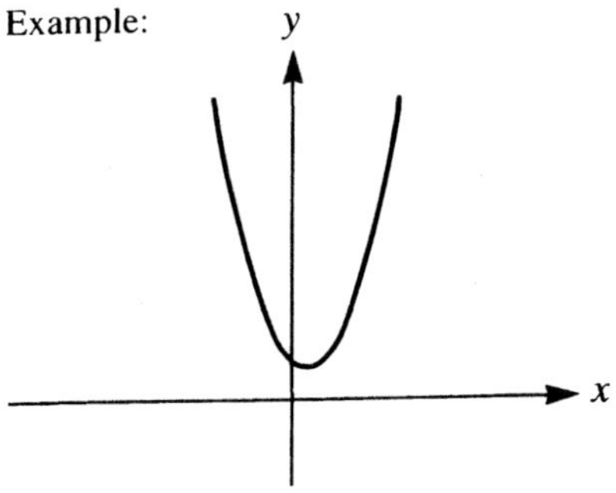

c. Example:

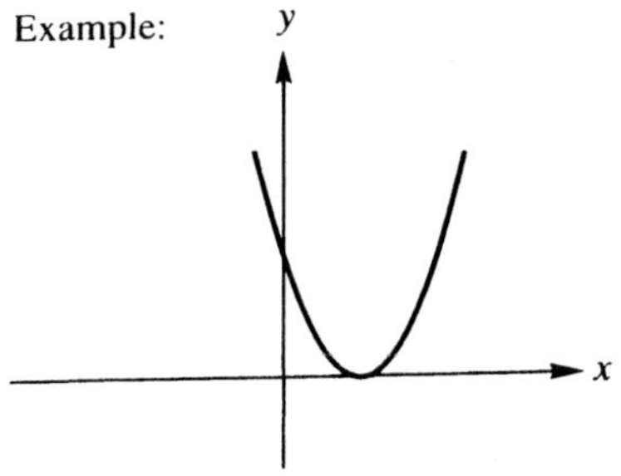

2.

2	3	−2	−5	−17	4	−11
		6	8	6	−22	−36
	3	4	3	−11	−18	−47

$\therefore \underline{\underline{P(2) = -47}}$

3. Yes. $P(x)$ could have zeros greater than 2 because the quotient and remainder (see Problem 2) has a sign reversal.

4.

Pos.	Neg.	Comp.
3	2	0
3	0	2
1	2	2
1	0	4

5.

$1 + i$	1	1	−4	7
		$1 + i$	$1 + 3i$	−6
	1	$2 + i$	$-3 + 3i$	1

$\therefore \underline{\underline{P(1 + i) = 1}}$

6. $(3, 10)(7, -2)$
$m = \dfrac{-2 - 10}{7 - 3} = -3$
$y - 10 = -3(x - 3)$
$\underline{\underline{3x + y = 19}}$

7. $P(x) = ax^3 + bx^2 + cx + d$
$(2, 16), (3, 30), (4, 36), (5, 28)$
$8a + 4b + 2c + d = 16$
$27a + 9b + 3c + d = 30$
$64a + 16b + 4c + d = 36$
$125a + 25b + 5c + d = 28$
Solving gives $a = -1, b = 5, c = 8, d = -12$
$\therefore \underline{\underline{P(x) = -x^3 + 5x^2 + 8x - 12}}$

8. *Airplane Payload Problem*
a. $P(L) = -L^3 + 5L^2 - 3L - 1$

L	−1	5	−3	−1, $P(L)$
−1	−1	6	−9	8
0	−1	5	−3	−1
1	−1	4	1	0
2	−1	3	3	5
3	−1	2	3	8
4	−1	1	1	3
5	−1	0	−3	−16

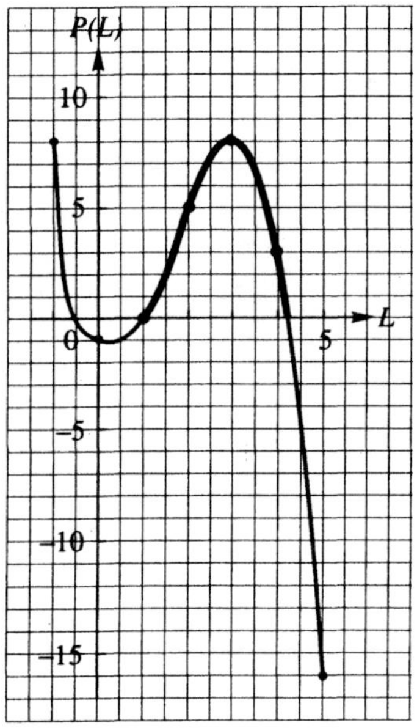

b. Integer zero is $\underline{\underline{1}}$.

c. $P(L) = -(L - 1)(L^2 - 4L - 1)$
$P(L) = 0$ if $L^2 - 4L - 1 = 0$
$$L = \frac{4 \pm \sqrt{16 - 4(1)(-1)}}{2}$$
$$L = \frac{4 \pm \sqrt{20}}{2} = 2 \pm \sqrt{5}$$
other zeros are $\underline{2 + \sqrt{5}, 2 - \sqrt{5}}$

d. See darkened portion of graph.

e. Longest plane is about $2 + \sqrt{5}$, or 4.236... tens of meters. About 42 meters

(*Note:* Problem 5 can count as extra credit.)

TEST 29 CHAPTER 10

1. $i^{37} = i^1 = \underline{i}$

2. $i^{-31} = \frac{1}{i^{31}} = \frac{1}{i^3} = \underline{i}$

3. $\sqrt{-12} \times \sqrt{-15}$
$= 2i\sqrt{3} \times i\sqrt{15}$
$= \underline{-6\sqrt{5}}$

4. $(3 + 9i) - (12 - 5i) = \underline{-9 + 14i}$

5. $\frac{8 + 6i}{7 - 5i} \cdot \frac{7 + 5i}{7 + 5i}$
$= \frac{56 + 82i - 30}{49 + 25}$
$= \frac{13 + 41i}{37} = \underline{\frac{13}{37} + \frac{41}{37}i}$

6. $x^2 - 6x + 13 = 0$
$$x = \frac{6 \pm \sqrt{36 - 4(1)(13)}}{2}$$
$$x = \frac{6 \pm \sqrt{-16}}{2}$$
$x = 3 \pm 2i$
$S = \{3 + 2i, 3 - 2i\}$

7. $9x^2 + 49 = \underline{(3x + 7i)(3x - 7i)}$

8. a. $3x^2 - 7x + 2 = 0$
$$x = \frac{7 \pm \sqrt{49 - 4(3)(2)}}{6}$$
$x = 2$ or $\frac{1}{3}$
$S = \left\{2, \frac{1}{3}\right\}$

b. Sum $= 2\frac{1}{3} = \frac{7}{3} = -\left(-\frac{7}{3}\right)$, which checks.

Product $= \frac{2}{3}$, which checks.

9. One solution is $-5 + 7i$.
So other solution is $-5 - 7i$.
Sum $= -10$; product $= 25 + 49 = 74$
$\therefore$ equation is $\underline{x^2 + 10x + 74 = 0}$

10. Example: $P(x) = x^4 - x^3 + x^2 - x + 1$, which has *four* sign reversals, and thus, potentially, four positive zeros.

11. $P(x) = ax^3 + bx^2 + cx + d$
$(-2, -35), (1, 7), (2, 9), (3, 25)$
$-8a + 4b - 2c + d = -35$
$a + b + c + d = 7$
$8a + 4b + 2c + d = 9$
$27a + 9b + 3c + d = 25$

Solving gives $a = 2, b = -5, c = 3, d = 7$
$\therefore$ equation is $\underline{P(x) = 2x^3 - 5x^2 + 3x + 7}$

12. Example:

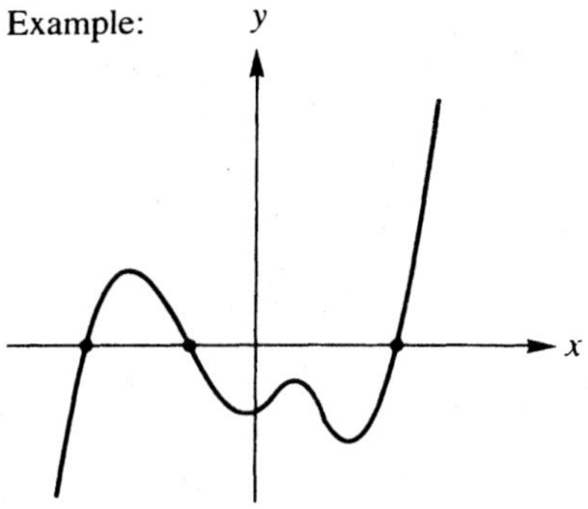

13. a.

x	1	-3	-5	10, $P(x)$
-3	1	-6	13	-29
-2	1	-5	-5	0
-1	1	-4	-1	11
0	1	-3	-5	10
1	1	-2	-7	3
2	1	-1	-7	-4
3	1	0	-5	-5
4	1	1	-1	6
5	1	2	5	35

b.

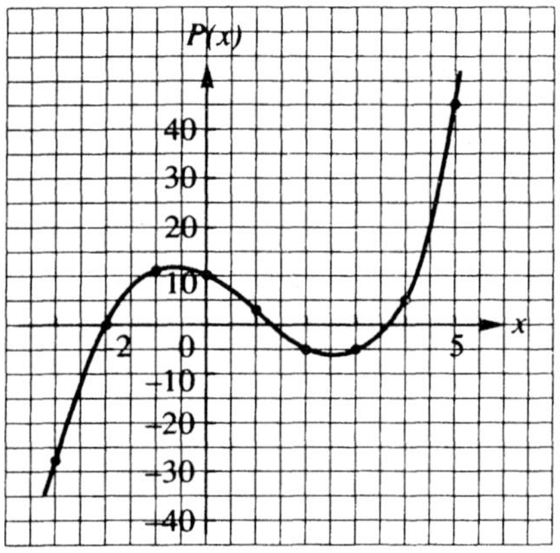

c. From chart, $P(-2) = 0$
$\therefore P(x) = (x + 2)(x^2 - 5x + 5)$
Other zeros come from:
$x^2 - 5x + 5 = 0$
$$x = \frac{5 \pm \sqrt{25 - 4(1)(5)}}{2}$$
$$x = \frac{5 \pm \sqrt{5}}{2}$$
Zeros are $-2, \frac{1}{2}(5 + \sqrt{5}), \frac{1}{2}(5 - \sqrt{5})$

or $\underline{-2, 3.618..., 1.381...}$

d. Since the quotient and remainder in the last row of the table in part (a) are all positive, the Upper Bound Theorem assures that 5 is an upper bound on the zeros of $P(x)$.

TEST 30 CHAPTER 11 Sections 11-1 to 11-6

1. a. 3, 8, 15, 24, 35, 48, 63, ...

① ② ③ ④ ⑤ ⑥ ⑦

1×3 2×4 3×5 4×6 5×7 6×8 7×9

$t_8 = 8 \times 10 = \underline{\underline{80}}$

$t_9 = 9 \times 11 = \underline{\underline{99}}$

b. $\underline{\underline{t_n = n(n + 2)}}$

2. a. $t_{73} = 4 + 72(3) = \underline{\underline{220}}$

b. $S_{73} = \dfrac{73}{2}(4 + 220) = \underline{\underline{8176}}$

c. $1075 = 4 + (n - 1)(3)$

$357 = n - 1$

$\underline{\underline{358 = n}}$

d. $23{,}001 = \dfrac{n}{2}(8 + (n - 1)(3))$

$3n^2 + 5n - 46{,}002 = 0$

$n = \dfrac{-5 \pm \sqrt{552{,}049}}{6}$

$n = 123$ or ~~$-124.66...$~~ no.

$\underline{\underline{n = 123}}$

3. a. $t_{15} = 83 \times 1.07^{14} = \underline{\underline{214.0183...}}$

b. $S_{15} = 83 \cdot \dfrac{1 - 1.07^{15}}{1 - 1.07}$

$= \underline{\underline{2085.708...}}$

c. $10{,}832 \approx 83 \times 1.07^{n-1}$

$n - 1 \approx \dfrac{\log\left(\dfrac{10832}{83}\right)}{\log 1.07}$

$n - 1 \approx 71.99995...$

$n = \underline{\underline{73}}$

d. $15408.15 \approx 83 \cdot \dfrac{1 - 0.07^n}{1 - 0.07}$

$13.9948... \approx 1.07^n$

$n \approx \dfrac{\log 13.9948...}{\log 1.07}$

$\underline{\underline{n = 39}}$

4. a. 4, __, __, __, 324

① ② ③ ④ ⑤

$b_5 = b_1 + 4d$

$324 = 4 + 4d$

$80 = d$

4, $\underline{\underline{84}}$, $\underline{\underline{164}}$, $\underline{\underline{244}}$, 324

b. $324 = 4 \cdot r^4$

$81 = r^4$

$r^2 = 9$ or $r^2 = -9$

$r = \pm 3$ or $r = \pm 3i$

4, $\underline{\underline{12}}$, $\underline{\underline{36}}$, $\underline{\underline{108}}$, 324

or: 4, $\underline{\underline{-12}}$, $\underline{\underline{36}}$, $\underline{\underline{-108}}$, 324

or: 4, $\underline{\underline{12i}}$, $\underline{\underline{-36}}$, $\underline{\underline{-108i}}$, 324

or: 4, $\underline{\underline{-12i}}$, $\underline{\underline{-36}}$, $\underline{\underline{108i}}$, 324

5. $\displaystyle\sum_{k=2}^{7} (-1)^k(k - 1)$

$= 1 - 2 + 3 - 4 + 5 - 6$

$= \underline{\underline{-3}}$

6. $200 - 170 + 144.5 - \ldots$

$r = -0.85$

$S = 200 \cdot \dfrac{1}{1 + 0.85}$

$= \underline{\underline{108.108...}}$

7. $0.324324324...$

$= 0.324 + 0.000324 + 0.000000324 + \ldots$

$S = 0.324 \cdot \dfrac{1}{1 - 0.001}$

$= \dfrac{324}{999} = \underline{\underline{\dfrac{12}{37}}}$

8. $t_1 = 10$; $r = 2i$

$10, 20i, -40, -80i, 160, 320i, -640, -1280i$

9.

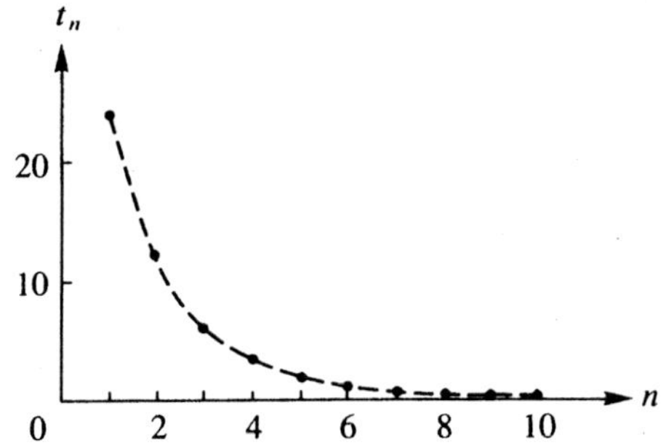

TEST 31 CHAPTER 11 Section 11-7

1. a. $t_1 = 5, t_2 = 15$, so $\underline{\underline{r = 3}}$.

$\underline{\underline{t_3 = 45}}$, $\underline{\underline{t_4 = 135}}$.

b. $t_1 = 5, t_2 = 15$, so $\underline{\underline{d = 10}}$.

$\underline{\underline{t_3 = 25}}$, $\underline{\underline{t_4 = 35}}$

2. $t_1 = 8, t_3 = 20$, so $\underline{\underline{t_5 = 32}}$.

$t_2 = \dfrac{1}{2}(8 + 20) = 14.$

$\therefore S_3 = 8 + 14 + 20 = \underline{\underline{42}}$

3. $t_1 = 7$ and $S = 70$. So:

$70 = \dfrac{7}{1 - r}$

$10 = \dfrac{1}{1 - r}$

$1 - r = 0.1$

$\underline{\underline{r = 0.9}}$

$\therefore S_2 = 7 + 6.3 = \underline{\underline{13.3}}$

4. $3 + 8 + 13 + 18 + 23 + \ldots$

$\underline{\underline{\text{Yes}}}$. 23 is the $\underline{\text{fifth}}$ term.

5. *Compound Interest Problem*

In 11.6 years there are 46 full quarters.

$\therefore M = 500(1.02^{46})$

$= \underline{\underline{\$1243.31}}$

6. *Oil Field Problem*

$t_1 = 80{,}000, t_2 = 70{,}000$

$\therefore r = \dfrac{7}{8}$

$S = 80{,}000 \cdot \dfrac{1}{1 - \dfrac{7}{8}}$

$= \underline{\underline{640{,}000 \text{ barrels}}}$

7. *Swing Problem*

a. Standing: $t_{20} = 3 \cdot (1.2^{20})$
$= 115.01\ldots$
About 115 cm

Sitting: $t_{20} = 3 + (20)(2)$
$= 43$ cm

b. Sitting: $75 = 3 + (n)(2)$
$n = 36$
36 pumps

Standing: $75 = 3 \cdot (1.2^n)$
$$n = \frac{\log\left(\frac{75}{3}\right)}{\log 1.2}$$
$n = 17.65\ldots$
18 pumps

c. If $t_0 = 0$, then $t_n = 0 \cdot (1.2^n)$ when standing.
So $t_n = 0$ for *all* values of n when you are standing.
When you are sitting, $t_n = 0 + (n)(2)$, and thus amplitudes are 0, 2, 4, … .

TEST 32 CHAPTER 11 Sections 11-8 to 11-10

1. a. $(r - p)^9 = r^9 - 9r^8p + 36r^7p^2 - 84r^6p^3 + 126r^5p^4 - 126r^4p^5 + 84r^3p^6 - 36r^2p^7 + 9rp^8 - p^9$

b. $(2x + y^3)^5$
$= (2x)^5 + 5(2x)^4(y^3) + 10(2x)^3(y^3)^2 + 10(2x)^2(y^3)^3 + 5(2x)(y^3)^4 + (y^3)^5$
$= 32x^5 + 80x^4y^3 + 80x^3y^6 + 40x^2y^9 + 10xy^{12} + y^{15}$

c. $(1 + i)^6 = 1^6 + 6(1^5)i + 15(1^4)(i^2) + 20(1^3)(i^3) + 15(1^2)(i^4) + 6(1)(i^5) + i^6$
$= 1 + 6i - 15 - 20i + 15 + 6i - 1$
$= -8i$

2. a. $(v - t)^{17}$: Term is $-\frac{17!}{12!\,5!}v^{12}t^5$

b. $(x + y)^{19}$: Term is $\frac{19!}{10!\,9!}x^{10}y^9$

c. $(p - j)^{55}$: Term is $\frac{55!}{49!\,6!}p^{49}j^6$

d. $(a + b^2)^{10}$: Term is $\frac{10!}{6!\,4!}a^6(b^2)^4$
$= \frac{10!}{6!\,4!}a^6b^8$

3. a. Row for 7th power of $(a + b)$.

b.
1 7 21 35 35 21 7 1
1 8 28 56 70 56 28 8 1 ← Ans.

4. a. $38 \cdot 37 \cdot 36 \cdot 35 \cdot 34 \cdot 33 = \frac{38!}{32!}$

b. $\frac{14!}{19 \cdot 18 \cdot 17 \cdot 16 \cdot 15} = \frac{14!\,14!}{19!}$

5. a. $\frac{8!}{4!} = 1680$ b. $\frac{11!}{3!\,8!} = 165$

c. $\frac{(n-1)!}{n!} = \frac{1}{n}$

6. The pattern is, for example, $3! = \frac{4!}{4}$.
Applying this gives $0! = \frac{1!}{1} = \frac{1}{1} = 1$.
$\therefore 0! = 1$ is a reasonable definition.

7. $n!$ is evenly divisible by 7 for all $n \geq 7$.

8. a. u = 7: Coefficient $= \frac{\text{(exp.)(coeff)}}{\text{(term no.)}} = \frac{(0)(1)}{6} = 0$
u = 8: Coefficient $= \frac{(-1)(0)}{(7)} = 0$

b. All following coefficients are zero, because each has the previous one.

TEST 33 CHAPTER 11

1. a. $13 + 19 + \ldots$; $d = 19 - 13 = 6$
$\therefore t_{27} = 13 + 26(6)$
$= 169$

b. $S_{27} = \frac{27}{2}(13 + 169)$
$= 2457$

c. $445 = 13 + (n - 1)(6)$
$\frac{445 - 13}{6} = n - 1$
$73 = n$

2. a. $50 + 48 + \ldots$; $r = \frac{48}{50} = 0.96$
$867.373 = 50 \cdot \frac{1 - 0.96^n}{1 - 0.96}$
$0.96^n = 0.306101\ldots$
$n = \frac{\log 0.306101\ldots}{\log 0.96}$
$n = 29$

b. $S = \frac{50}{1 - 0.96}$
$= 1250$

3. 35, ___, ___, ___, ___, 98
① ② ③ ④ ⑤ ⑥
$98 = 35 + (6 - 1)(d)$
$12.6 = d$
35, 47.6, 60.2, 72.8, 85.4, 98

4. $\sum_{k=1}^{5} 2^k + 3 = 5 + 7 + 11 + 19 + 35$
$= 77$

b. $7 - 5 = 2$ and $11 - 7 = 4$, *not* arithmetic
$7 \div 5 \neq 11 \div 7$, not geometric
neither

5. a. 0, 3, 8, 15, 35, …
① ② ③ ④ ⑤
$t_6 = 48$, $t_7 = 63$

b. $t_n = n^2 - 1$ or $t_n = (n + 1)(n - 1)$

6. $(p^2 + 3h)^4 = (p^2)^4 + 4(p^2)^3(3h) + 6(p^2)^2(3h)^2 + 4(p^2)(3h)^3 + (3h)^4$
$= p^8 + 12p^6h + 54p^4h^2 + 108p^2h^3 + 81h^4$

7. a. $(a - b)^9$: term $= -\frac{9!}{4!\,5!}a^4b^5 = \underline{\underline{-126a^4b^5}}$

b. $(x^{\frac{1}{2}} + x^{-\frac{1}{2}})^8$

7th term has $(x^{-\frac{1}{2}})^6$.

$\therefore$ term is $\frac{8!}{2!\,6!}(x^{\frac{1}{2}})^2(x^{-\frac{1}{2}})^6$

$= \underline{\underline{28x^{-2}}}$

8. *Office Building Problem*

a. Let n = no. of stories

Let t_n = \$/m² for nth story.

$t_n = 500(1.2^{n-1})$

$t_{10} = 500(1.2^9) = 2579.89\ldots$

About \$2580 per m².

b. Costs are 500,000 + 600,000 + ...

$\therefore S_{10} = 500{,}000 \cdot \frac{1 - 1.2^{10}}{1 - 1.2}$

$= 12{,}979{,}341$

About \$13,000,000

c. $500 \times 1.2^{n-1} \le 9000$

$n - 1 \le \frac{\log \frac{9000}{500}}{\log 1.2}$

$n \le 16.85\ldots$

Max of 16 stories

9. Example:

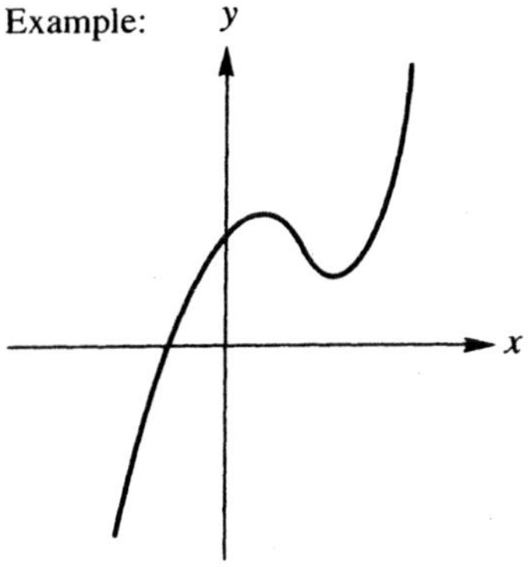

TEST 34 CHAPTER 12 Sections 12-1 to 12-5

1. Permutations are involved if the *order* in which elements are arranged makes a difference. Combinations are involved if only the *elements* in the group make a difference, not the order in which they are arranged.

2. Outcomes are *equally likely results* of a random experiment. Events are *sets* of outcomes.

3. $n = {}_8C_3 = 56$ 4. $n = {}_8P_3 = 336$

5. ${}_9C_6 = 84$ 6. ${}_{11}P_4 = 7920$

7. ${}_5C_5 = 1$

8. a. $n = {}_9P_4 = 3024$ b. $n = 9! = 362{,}880$

c. $n = 25{,}200$ d. Fixed position

e. $P = \frac{5}{72}$

9. a. $n = {}_{20}C_7 = 77{,}520$

b. $n = {}_{12}C_3 \cdot {}_8C_4 = 15{,}400$

c. $n = {}_{12}C_3 \cdot {}_8C_4 + {}_{12}C_4 \cdot {}_8C_3 = 43{,}120$

d. $P = \frac{539}{969}$ e. $n = {}_7C_5 = 21$

f. $n = {}_7P_5 = 2520$

TEST 35 CHAPTER 12

1. The sample space is the set of all outcomes.

2. $P = \frac{7}{13}$ 3. $n = {}_{13}P_9 = 259{,}459{,}200$

4. $n = {}_{30}C_{11} = 54{,}627{,}300$

5.

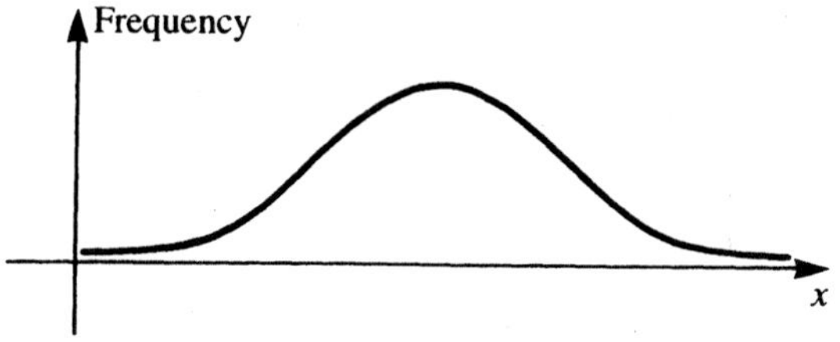

6. *Sawmill Problem*

a. i. 0.98

ii. 0.97

iii. 0.9506

iv. 0.0006

b.

Event	P	Payoff	$P \times$ Payoff
Arm only	0.0194	300	5.82
Leg only	0.0294	400	11.76
Both	0.0006	7000	4.20
Neither	0.9506	0	0.00
	1.0000		21.78

Expectation is \$21.78 per policy.

c. \$26.78

7. a. 0.4

b.

x	$P(x)$
0	0.01024
1	0.0768
2	0.2304
3	0.3456
4	0.2592
5	0.07776
c. $\longrightarrow$	1.00000

d. P(at least 3) $= 0.68256$

e. Binomial distribution, because the probabilities are terms in the binomial expansion of $(0.4 + 0.6)^5$.

f.

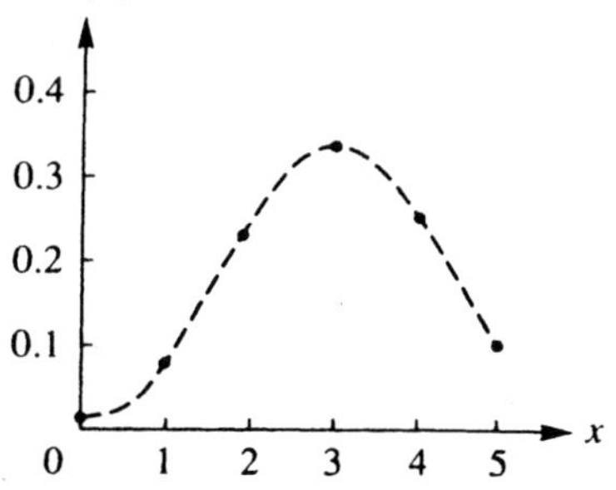

8. a.

x	dev	dev^2
93	5.2	27.04
82	−5.8	33.64
95	7.2	51.84
86	−1.8	3.24
90	2.2	4.84
84	−3.8	14.44
89	1.2	1.44
79	−8.8	77.44
87	−0.8	0.64
93	5.2	27.04
878		241.6

$\text{Mean} = \dfrac{878}{10} = 87.8$

$\text{Std. dev.} = \sqrt{\dfrac{241.6}{9}}$

$= \underline{\underline{5.181\ldots}}$

b. i. Random differences in the abilities of students in the other classes.
ii. If other classes have many more than 10 students, performance could be lower.
iii. If later classes found out what was on the test, performance could be higher.
iv. Later classes sometimes perform lower because students are tired late in the day.

TEST 36 CHAPTERS 9 through 12

1. $\left(\dfrac{x+3}{1}\right)^2 + \left(\dfrac{y-2}{4}\right)^2 < 1$

$16(x^2 + 6x + 9) + (y^2 - 4y + 4) < 16$

$\underline{\underline{16x^2 + y^2 + 96x - 4y + 132 < 0}}$

2. $4x - 9y^2 - 16x - 18y = 29$

$4(x^2 - 4x + 4) - 9(y^2 + 2y + 1) = 36$

$\left(\dfrac{x-2}{3}\right)^2 - \left(\dfrac{y+1}{2}\right)^2 = 1$

3. $16x^2 + 3y^2 - 32x + 6y - 72 = 0$

$2x + y = 8 \longrightarrow y = 8 - 2x$

$16x^2 + 3(64 - 32x + 4x^2) - 32x + 6(8 - 2x) - 72 = 0$

$28x^2 - 140x + 168 = 0$

$x^2 - 5x + 6 = 0$

$(x - 3)(x - 2) = 0$

$x = 3$ or $x = 2$

$y = 8 - 2(3)$ $\quad$ $y = 8 - 2(2)$

$y = 2$ $\quad$ $y = 4$

$\underline{\underline{S = \{(3, 2), (2, 4)\}}}$

4. $d_1 = d_2$

$|x - 3| = \sqrt{(x-5)^2 + (y-2)^2}$

$x^2 - 6x + 9 = x^2 - 10x + 25 + y^2 - 4y + 4$

$4x = y^2 - 4y + 20$

The graph is a parabola.

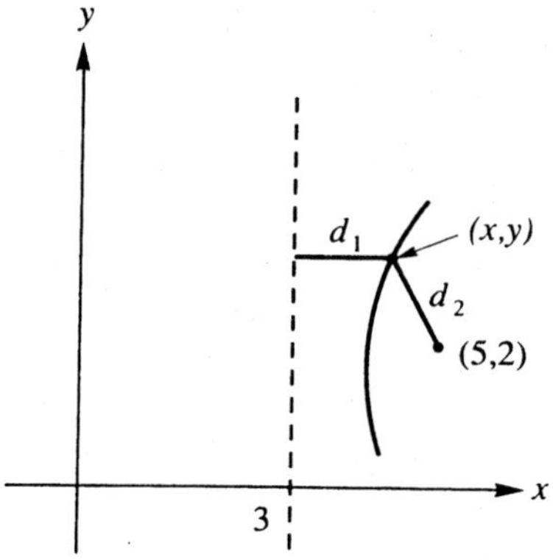

5. $\underline{\underline{\text{5th degree}}}$, $\underline{\underline{\text{3 real zeros}}}$, $\underline{\underline{\text{2 non-real complex zeros}}}$

6. $f(x) = x^3 - 7x^2 + 17x - 15$

a.

4⌋	1	−7	17	−15
		4	−12	20
	1	−3	5	5

$\underline{\underline{f(4) = 5}}$

b.

3⌋	1	−7	17	−15
		3	−12	15
	1	−4	5	0

$\therefore f(3) = 0,$ Q.E.D.

c. $f(x) = (x - 3)(x^2 - 4x + 5)$

$f(x) = 0 \Leftrightarrow x = 3$ or $x^2 - 4x + 5 = 0$

$x = 2 \pm i$

Other zeros are $\underline{\underline{2 + i}}$ and $\underline{\underline{2 - i}}$.

d. $f(x) = x^3 - 7x^2 + 17x - 15$ has 3 sign reversals. So the number of positive zeros must be 3 or 1. The fact that there is exactly $\underline{\text{one}}$ positive zero confirms Descartes' rule of signs.

e.

8⌋	1	−7	17	−15
		8	8	200
	1	1	25	185

The quotient and remainder have no sign reversals. So 8 is an upper bound on the zeros of $f(x)$.

7. $i^{-23} = i^{-3} = \underline{\underline{i}}$

8. $\dfrac{50}{3+4i} \cdot \dfrac{3-4i}{3-4i}$

$= \dfrac{50(3-4i)}{9+16}$

$= \underline{\underline{6 - 8i}}$

9. One solution is $2 + 3i$.

$\therefore$ other is $2 - 3i$

Sum $= 4$, product $= 4 + 9 = 13$

$\therefore$ equation is $\underline{\underline{x^2 - 4x + 13 = 0}}$

10. $P(x) = ax^3 + bx^2 + cx + d$

(3, 42), (2, 14), (1, 4), (0, 0).

$27a + 9b + 3c + d = 42$

$8a + 4b + 2c + d = 14$
$a + b + c + d = 4$
$0 + 0 + 0 + d = 0$
$\therefore d = 0$
Solving gives $a = 2, b = -3, c = 5$.
$\therefore$ equation is $P(x) = 2x^3 - 3x^2 + 5x$

11. 1, 3, 7, 15, 31, 63, 127, 255
① ② ③ ④ ⑤ ⑥ ⑦ ⑧
$t_n = 2^n - 1$

12. 5, 9, 13, ... ; $d = 4$
$t_{71} = 5 + 70(4)$
$= 285$

13. $S_{17} = 81 \times \dfrac{1 - 1.09^{17}}{1 - 1.09}$
$= 2994.870\ldots$

14. 6, __, __, __, 486
① ② ③ ④ ⑤
$486 = 6 + (5 - 1)(d)$
$\dfrac{486 - 6}{4} = d$
$120 = d$
6, 126, 246, 366, 486

15. *Nail Problem*
a. 0.8, 0.7, ...
$r = \dfrac{0.7}{0.8}$
$= 0.875$
$\therefore S = \dfrac{0.8}{1 - 0.875}$
$= 6.4$ cm

b.
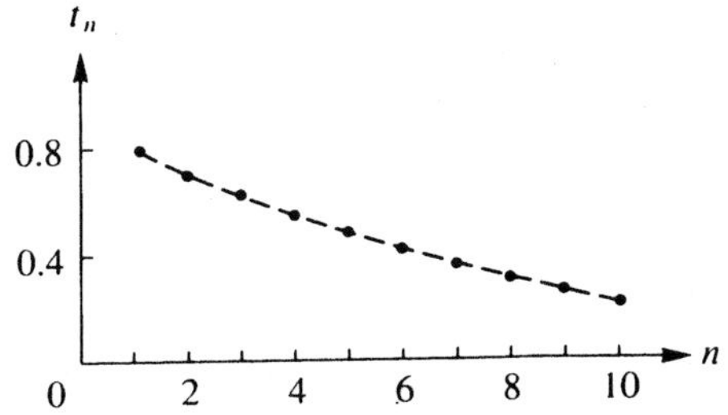

16. Term is $-\dfrac{8!}{3!\,5!}a^3(b^2)^5$
$= -56a^3b^{10}$

17. $P(\text{black face}) = \dfrac{6}{52} = \dfrac{3}{26}$

18. ${}_{12}P_5 = 12 \cdot 11 \cdot 10 \cdot 9 \cdot 8$
$= 95{,}040$

19. $3 \cdot 9 \cdot 8 \cdot 7 \cdot 6 \cdot 5 \cdot 4 \cdot 3 \cdot 2 \cdot 1$
$= 1{,}088{,}640$

20. no. of committees $= {}_{10}C_3 \cdot {}_9C_4$
$= \dfrac{10!}{3!\,7!} \cdot \dfrac{9!}{4!\,5!}$
$= 120 \cdot 126$
$= 15{,}120$

21. $P(\text{at least one wins})$
$= 1 - P(\text{both lose})$
$= 1 - (0.2)(0.3)$
$= 0.94$
94%

22.

	Prob.	Payoff	Total
$P(3) = {}_3C_3(0.6)^3$	0.216	50	10.80
$P(2) = {}_3C_2(0.6)^2(0.4)$	0.432	10	4.32
$P(1) = {}_3C_1(0.6)(0.4)^2$	0.288	3	0.864
$P(0) = {}_3C_0(0.4)^3$	0.064	−20	−1.28
			Total: 14.704

About $14.70

23.

x	dev.	dev.^{2}
57	6	36
43	−8	64
49	−2	4
51	0	0
55	4	16
255		120

$\bar{x} = \dfrac{255}{5} = 51$
Std. dev. $= \sqrt{\dfrac{120}{4}}$
$= 5.477225\ldots$

TEST 37 CHAPTER 13 Sections 13-1 to 13-4

1. Periodic function

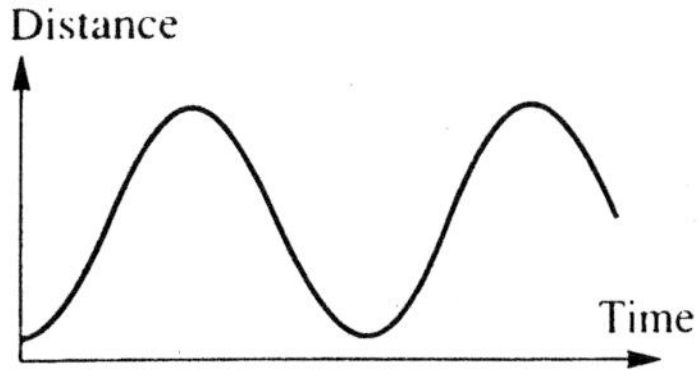

2.

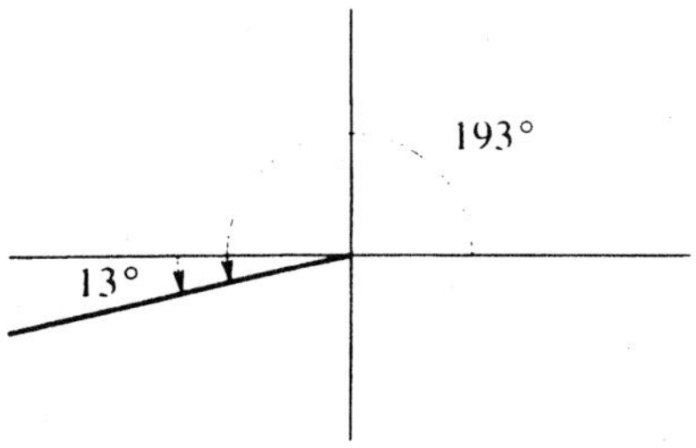

3.

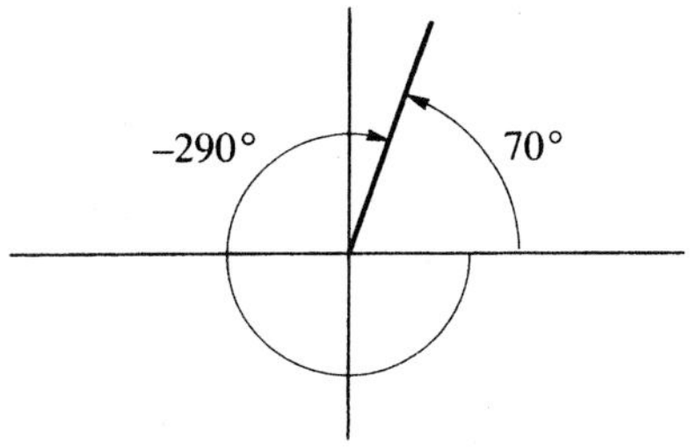

4.

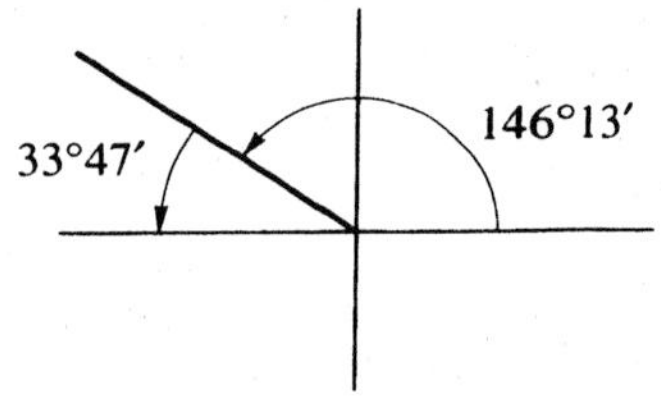

5.

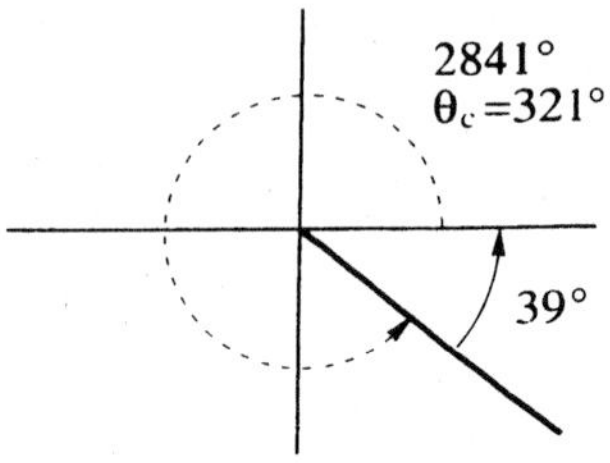

6. $x = \frac{2\pi}{3}$; $\theta = 120°$

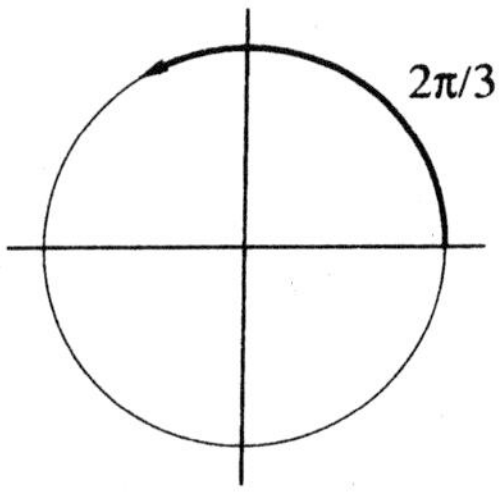

7. $\frac{\pi}{5}$; $\theta = 36°$

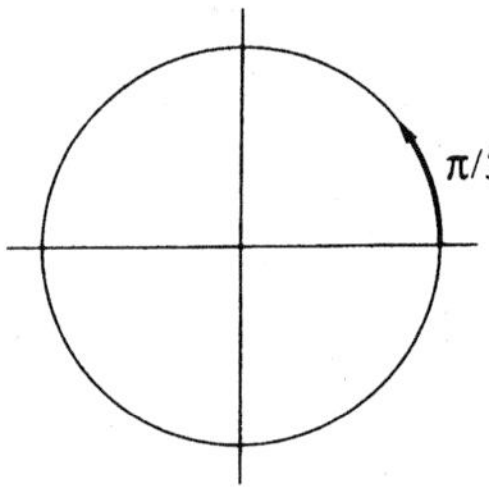

8. $x = \pi$; $\theta = 180°$

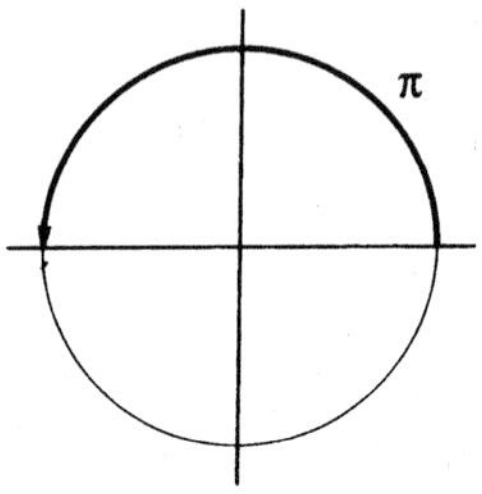

9. $x = 2$; $\theta = 114.59\ldots°$

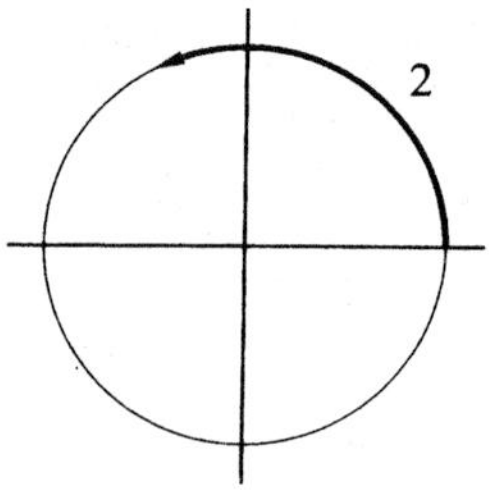

10. $\sin(-120°) = -\frac{\sqrt{3}}{2}$

$\cos(-120°) = -\frac{1}{2}$

$\tan(-120°) = \sqrt{3}$

$\cot(-120°) = \frac{1}{\sqrt{3}}$

$\sec(-120°) = -2$

$\csc(-120°) = -\frac{2}{\sqrt{3}}$

11. $\sec 180° = -1$

12. $\cot 6360° = \cot 240° = \cot 60° = \frac{\sqrt{3}}{3}$

13. $\sin^2 60° = \left(\frac{\sqrt{3}}{2}\right)^2 = \frac{3}{4}$

14. $\cos \pi = -1$

15. $\tan \frac{3\pi}{4} = -1$

16. csc 0: no value

17. $\sin 217° \approx -0.6018$

18. $\tan 24°37' \approx 0.4582$

19. $\sec 82° \approx 7.185$

20. $\cos 0.345 \approx 0.9411$

21. $\cot 1 \approx 0.6421$

22. $\csc 2 \approx 1.100$

23. $m^R(150°) = 150\left(\frac{\pi}{180}\right) = \frac{5\pi}{6}$

24. $\theta = 150°$

25.

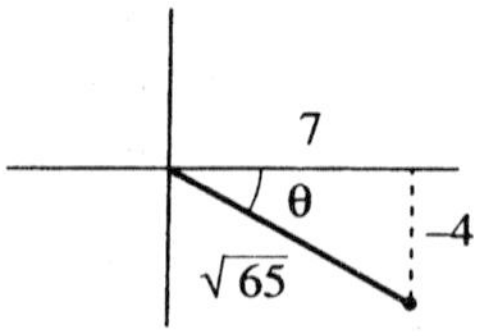

$\sin \theta = -\frac{4}{\sqrt{65}}$

26. Linear: $y = mx + b$

27. Quadratic: $y = ax^2 + bx + c$

1. $y = 10 + 6 \sin 5(\theta - 10°)$

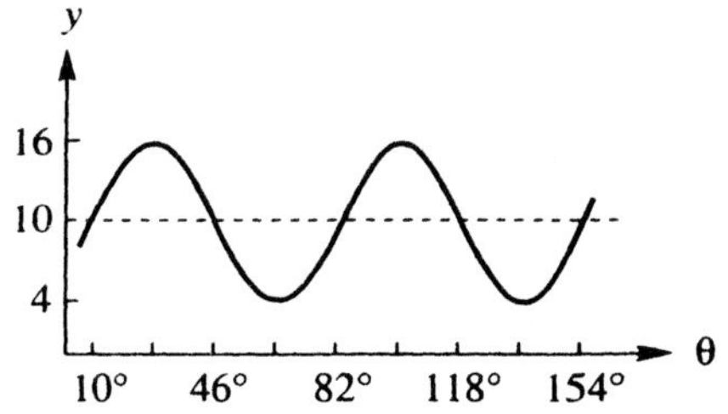

2. $y = \tan \theta$

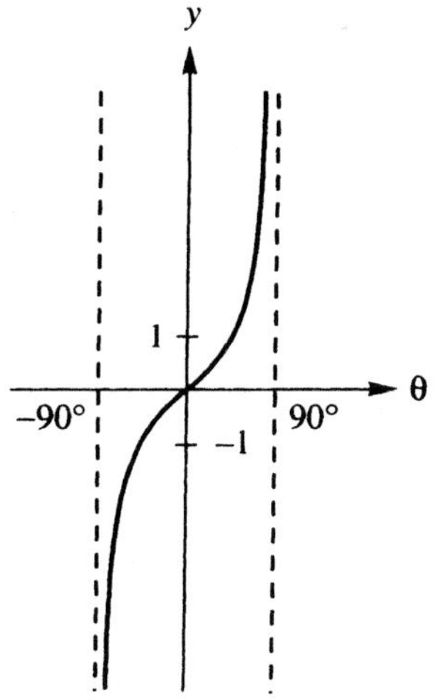

3. $y = \sec x$

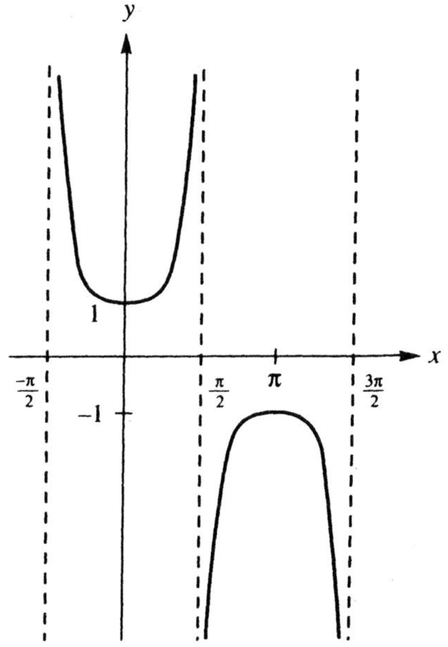

4. $y = 2 + 9 \cos \frac{\pi}{22}(x + 8)$

5. a.

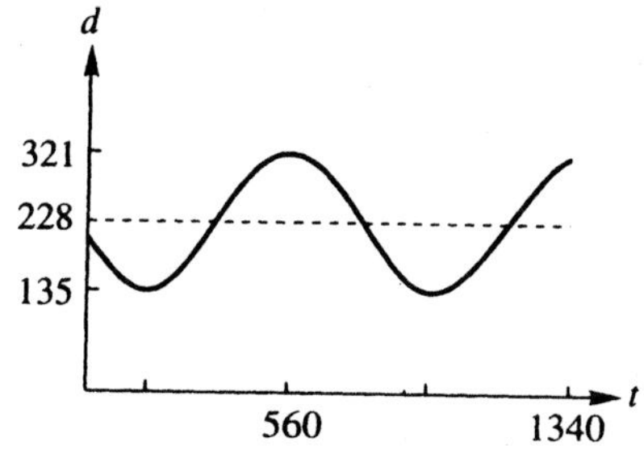

b. 209 million miles

c. Farthest: 321 million miles
Closest: 135 million miles

d. 780 days (the period)

e. (Bonus!)

$$\frac{1}{\text{period Earth}} - \frac{1}{\text{period Mars}} = \frac{1}{\text{period}}$$

$$\therefore \frac{1}{365} - \frac{1}{\text{period Mars}} = \frac{1}{780}$$

period Mars = 686 days
(Actual period is 687 days.)

6. $m^R(270°) = \frac{3\pi}{2}$

7. $m°\left(\frac{2\pi}{3}\right) = 120°$

8. $\cos \frac{\pi}{4} = \frac{\sqrt{2}}{2}$

9. $\sin \frac{4\pi}{3} = \frac{\sqrt{3}}{-2}$

10. $\cot \frac{\pi}{2} = 0$

11. $\tan 120° = -\sqrt{3}$

12. $\csc 135° = \sqrt{2}$

13. $\sec 180° = -1$

14. $\sec \theta = -\frac{\sqrt{13}}{2}$

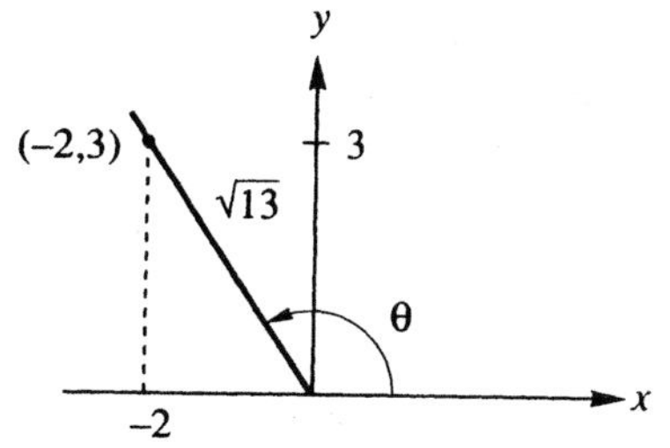

15. $\theta = 3504°$
$\theta_c = 264°$
$\theta_{ref.} = 84°$

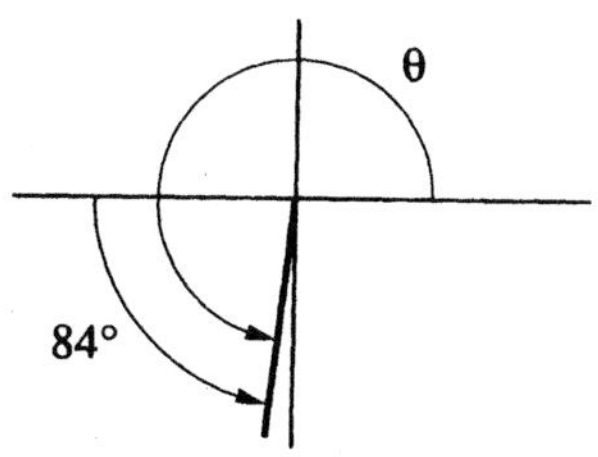

TEST 39 CHAPTER 13 Sections 13-9 to 13-11

1. $\text{Cos}^{-1}\dfrac{\sqrt{3}}{2} = \underline{\underline{\dfrac{\pi}{6}}}$

2. $\arcsin\dfrac{1}{2} = \underline{\underline{\dfrac{\pi}{6} + 2\pi n \quad \text{or} \quad \dfrac{5\pi}{6} + 2\pi n}}$

3. $\text{Arctan}\,(-1) = \underline{\underline{-\dfrac{\pi}{4}}}$

4. $\text{Arccot}\,(-1) = \underline{\underline{\dfrac{3\pi}{4}}}$

5. $\text{Sec}^{-1}2 = \underline{\underline{\dfrac{\pi}{3}}}$

6. $\text{Sin}^{-1}2$ $\underline{\underline{\text{no value}}}$

7. $\tan\left(\text{Cos}^{-1}\dfrac{12}{13}\right) = \underline{\underline{\dfrac{5}{12}}}$

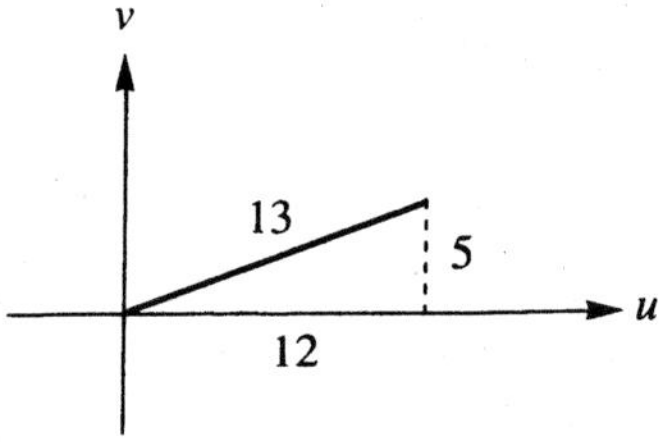

8. $\csc\,(\text{Arccsc}\,\sqrt{31}) = \underline{\underline{\sqrt{31}}}$

9. $y = 6 + 8\cos\dfrac{\pi}{7}(x - 4)$

   a. $x = 17$: $\underline{y = 13.207\ldots}$

   b. $x = 4 + \dfrac{7}{\pi}\cos^{-1}\left(\dfrac{y-6}{8}\right)$

   c. $y = 11$: $\underline{\underline{x = 2.0043\ldots,\ 5.9956\ldots,\ 16.0043\ldots}}$

10. a.

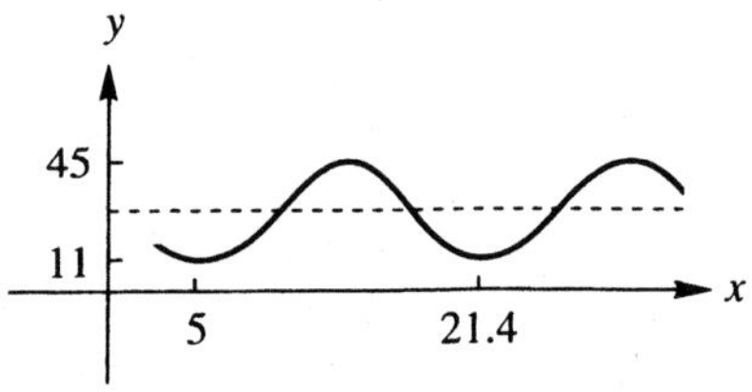

   b. $y = 28 - 17\cos\dfrac{\pi}{8.2}(x - 5)$

   c. $y = 43$:
   $x = 5 \pm 6.92\ldots + 16.4n$
   $x = 11.92\ldots + 16.4n$
   or $\quad -1.92\ldots + 16.4n$
   $x = 11.92\ldots,\ 14.47\ldots,\ \ldots$
   Second time is $\underline{\underline{14.5 \text{ sec.}}}$

11. $y = \cos^{-1}x$

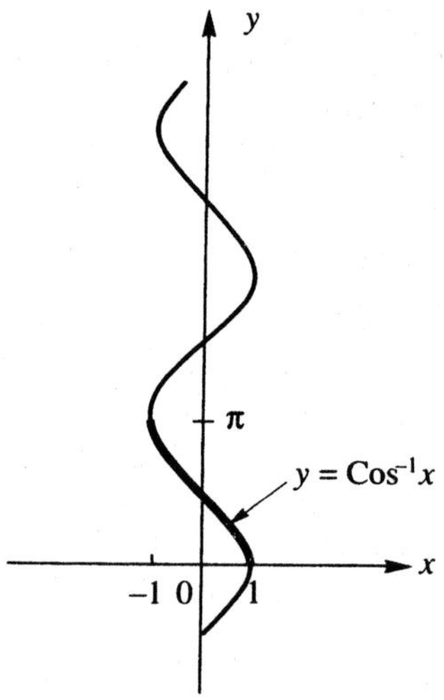

12. $y = \tan x$ and $y = \text{Arctan}\,x$ are reflections of each other through the line $y = x$.

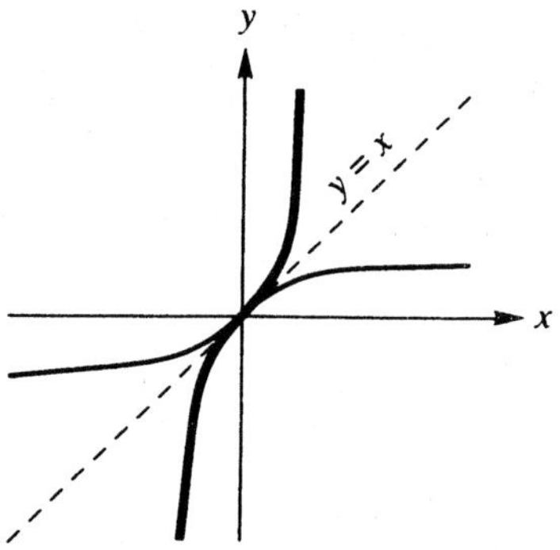

TEST 40 CHAPTER 13

1. $\cos 45° = \underline{\underline{\dfrac{\sqrt{2}}{2}}}$

2. $\cot\dfrac{2\pi}{3} = \underline{\underline{-\dfrac{\sqrt{3}}{3}}}$

3. $\text{Sec}^{-1}(-\sqrt{2}) = \underline{\underline{\dfrac{3\pi}{4}}}$

4. $\sin 63°19' = 0.89350\ldots$

5. $\csc 1.2 = \underline{1.0729\ldots}$

6. $\text{Arctan}\,3 = \underline{\underline{1.2490\ldots}}$

7. $\underline{\underline{0°}}$ because $\cot 0°$ is undefined.

8. $\underline{\underline{\dfrac{\pi}{6}}}$ because $\cos\dfrac{\pi}{6} = \dfrac{\sqrt{3}}{2}$.

9. $y = \tan x$

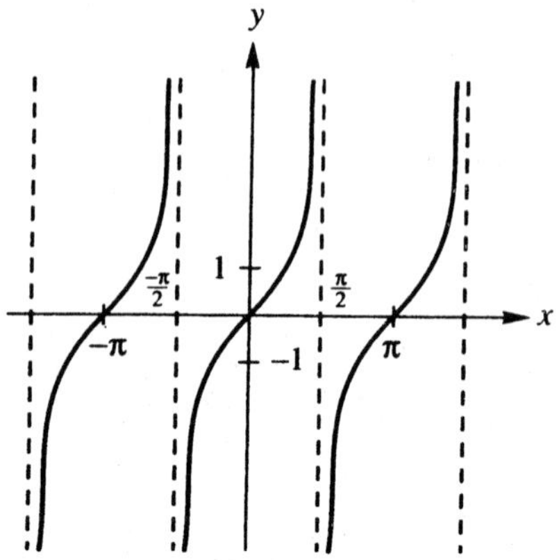

10. $y = \sin^{-1} x$

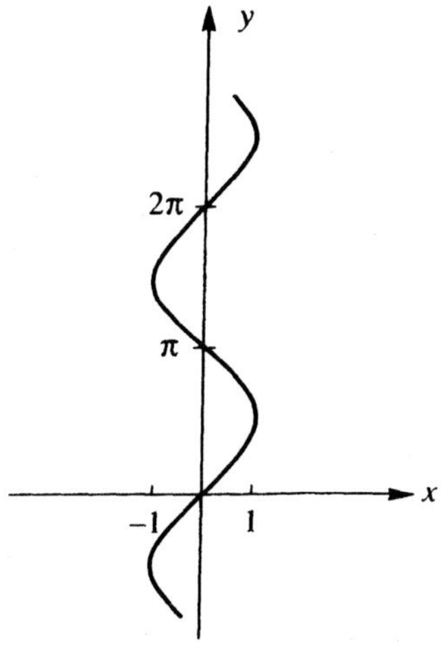

11. $y = 51 + 29 \cos 4.5\,(\theta - 70°)$

12. *Breathing Problem*

a. Let V = no. of liters.
Let t = no. of seconds.

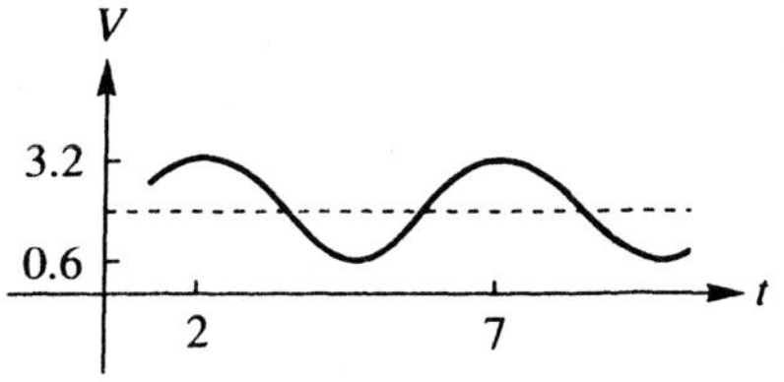

b. $V = 1.9 + 1.3 \cos \frac{2\pi}{5}(t - 2)$

c. $t = 0$: $V = 0.84827...$
About 0.85 liters

d. $V = 1$:

$1.9 + 1.3 \cos \frac{2\pi}{5}(t - 2) = 1$

$t = 2 \pm 1.85851... + 5n$

$t = 3.85851... + 5n$ or $0.14148... + 5n$

Between 0 and 0.14 seconds.

TEST 41 CHAPTER 14 Sections 14-1 and 14-2

1. $(1 - \tan A)^2$
$= 1 - 2 \tan A + \tan^2 A$
$= \sec^2 A - 2 \tan A.$
$\therefore (1 - \tan A)^2 = \sec^2 A - 2 \tan A$, Q.E.D.

2. $\cos^3 B - \cos^5 B$
$= \cos^3 B(1 - \cos^2 B)$
$= \cos^3 B \sin^2 B$
$\therefore \cos^3 B \sin^2 B = \cos^3 B - \cos^5 B$, Q.E.D.

3. $\frac{1}{1 - \cos C} + \frac{1}{1 + \cos C}$

$= \frac{1 + \cos C + 1 - \cos C}{1 - \cos^2 C}$

$= \frac{2}{\sin^2 C}$

$= 2\csc^2 C$

$\therefore \frac{1}{1 - \cos C} + \frac{1}{1 + \cos C} = 2 \csc^2 C$, Q.E.D.

4. $\csc D \cos^2 D + \sin D$
$= \csc D(\cos^2 D + \sin^2 D)$
$= \csc D$
$\therefore \csc D \cos^2 D + \sin D = \csc D$, Q.E.D.

5. $(5 \cos E - 2 \sin E)^2 + (2 \cos E + 5 \sin E)^2$
$= 25 \cos^2 E - 20 \cos E \sin E + 4 \sin^2 E$
$+ 4 \cos^2 E + 20 \cos E \sin E + 25 \sin^2 E$
$= 29 \cos^2 E + 29 \sin^2 E$
$= 29$
$\therefore (5 \cos E - 2 \sin E)^2$
$+ (2 \cos E + 5 \sin E)^2 = 29$, Q.E.D.

6. $\frac{\csc F}{\cos F} - \frac{\cos F}{\sin F}$

$= \frac{\csc F \sin F - \cos^2 F}{\cos F \sin F}$

$= \frac{1 - \cos^2 F}{\cos F \sin F}$

$= \frac{\sin^2 F}{\cos F \sin F}$

$= \tan F$

$\therefore \frac{\csc F}{\cos F} - \frac{\cos F}{\sin F} = \tan F$, Q.E.D.

TEST 42 CHAPTER 14, Sections 14-3 and 14-4

1. $\sin(-x) = -\sin x$ **2.** $\cos(-x) = \cos x$

3. $\tan\left(\frac{\pi}{2} - x\right) = \cot x$

or $\tan(90° - \theta) = \cot \theta$

4. $\cos 2x = 2\left(-\frac{2}{3}\right)^2 - 1$

$= -\frac{1}{9}$

$\sin 2x = 2\left(-\frac{2}{3}\right)\left(\frac{\sqrt{5}}{3}\right)$

$= -\frac{4\sqrt{5}}{9}$

$\tan 2x = \frac{\left(-\frac{4\sqrt{5}}{9}\right)}{\left(-\frac{1}{9}\right)}$

$= 4\sqrt{5}$

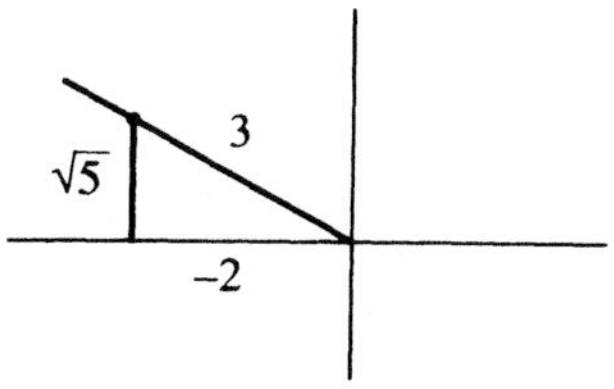

5. $\tan 12x = \frac{2 \tan 6x}{1 - \tan^2 6x}$

6. $\cos 3x = \cos(2x + x)$
$= \cos 2x \cos x - \sin 2x \sin x$
$= (2\cos^2 x - 1) \cos x - 2 \sin^2 x \cos x$
$= 2 \cos^3 x - 2 \cos x - 2(1 - \cos^2 x)\cos x$
$\therefore \cos 3x = 4 \cos^3 x - 3 \cos x$

7. $\tan 15° = \tan(60° - 45°)$
$= \dfrac{\tan 60° - \tan 45°}{1 + \tan 60° \tan 45°}$
$= \dfrac{\sqrt{3} - 1}{1 + \sqrt{3} \cdot 1}$
$= \underline{\underline{2 - \sqrt{3}}}$

8. $\sin(\theta + 60°) + \cos(\theta + 30°)$
$= \sin\theta\cos 60° + \cos\theta\sin 60° + \cos\theta\cos 30° - \sin\theta\sin 30°$
$= \sin\theta \cdot \left(\dfrac{1}{2}\right) + \cos\theta\left(\dfrac{\sqrt{3}}{2}\right) - \sin\theta\left(\dfrac{1}{2}\right) + \cos\theta\left(\dfrac{\sqrt{3}}{2}\right)$
$= \sqrt{3}\cos\theta.$
$\therefore \sin(\theta + 60°) + \cos(\theta + 30°) = \sqrt{3}\cos\theta,$
Q.E.D.

9. $\cos^2 x - \sin^2 x = \dfrac{\sqrt{3}}{2}$
$\cos 2x = \dfrac{\sqrt{3}}{2}$
$2x = \pm\dfrac{\pi}{6} + 2\pi n$
$S = \left\{\dfrac{\pi}{12}, \dfrac{11\pi}{12}, \dfrac{13\pi}{12}, \dfrac{23\pi}{12}\right\}$

10. $p = 30 + 17\cos\dfrac{\pi}{12}(f - 7)$

TEST 43 CHAPTER 14 Sections 14-5 to 14-8

1. $\sin 2A$
$= 2\left(-\dfrac{5}{13}\right)\left(\dfrac{12}{13}\right)$
$= \underline{\underline{-\dfrac{120}{169}}}$

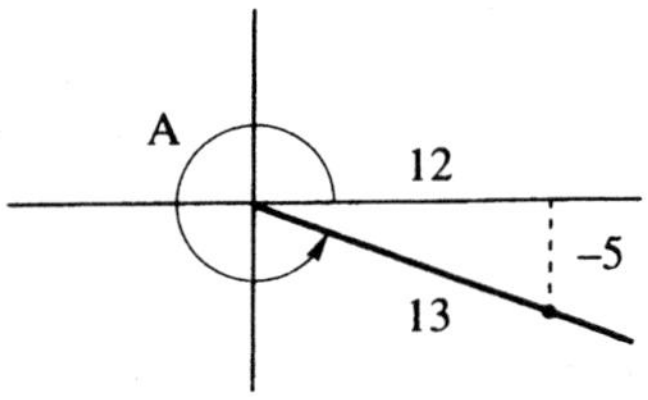

2. $\cos 2A = 2\left(\dfrac{12}{13}\right)^2 - 1$
$= \underline{\underline{\dfrac{119}{169}}}$

3. $\sin\dfrac{1}{2}A = +\sqrt{\dfrac{1}{2}\left(1 - \dfrac{12}{13}\right)}$
$= \underline{\underline{\dfrac{1}{\sqrt{26}}}}$

4. $\cos\dfrac{1}{2}A = -\sqrt{\dfrac{1}{2}\left(1 + \dfrac{12}{13}\right)}$
$= \underline{\underline{-\dfrac{5}{\sqrt{26}}}}$

5. $\cos 2x = \cos^2 x - \sin^2 x$

6. $\cos 2x = 2\cos^2 x - 1$

7. $\cos 2x = 1 - 2\sin^2 x$

8. $\tan\dfrac{1}{2}x = \pm\sqrt{\dfrac{1 - \cos x}{1 + \cos x}}$

9. $\tan\dfrac{1}{2}x = \dfrac{1 - \cos x}{\sin x}$ or $\tan\dfrac{1}{2}x = \dfrac{\sin x}{1 + \cos x}$

10. $\cos 37° \cos 99° = \dfrac{1}{2}\cos 136° + \dfrac{1}{2}\cos 62°$

11. $\sin 4 + \sin 7 = 2\sin 5.5\cos 1.5$

12. $\cos 8x - \cos 2x = -2\sin 5x\sin 3x$

13. $\cos 5x\sin 9x = \dfrac{1}{2}\sin 14x + \dfrac{1}{2}\sin 4x$

14. $19\cos(\theta - 50°) = 12.21\ldots\cos\theta + 14.55\ldots\sin\theta$

15. $8\cos\theta + 15\sin\theta = 17\cos(\theta - 61.92\ldots°)$

16. $\dfrac{1 - \cos\theta}{\sin\theta} = \sqrt{3},\ 0° \le \theta < 360°$
$\tan\dfrac{1}{2}\theta = \sqrt{3}$
$\dfrac{1}{2}\theta = 60° + 180n°$
$\theta = 120° + 360n°$
$S = \{120°\}$

17. $\cos 4x + \cos 2x = 0, \quad 0 \le x < \pi$
$2\cos 3x\cos x = 0$
$\cos 3x = 0$ or $\cos x = 0$
$3x = \dfrac{\pi}{2} + \pi n$ or $x = \dfrac{\pi}{2} + \pi n$
$x = \dfrac{\pi}{6} + \left(\dfrac{\pi}{3}\right)n$ *or* $x = \dfrac{\pi}{2} + \pi n$
$S = \left\{\dfrac{\pi}{6}, \dfrac{5\pi}{6}, \dfrac{\pi}{2}\right\}$

TEST 44 CHAPTER 14

1. $\cos^2 x = 1 - \sin^2 x$

2. $\cos 2x = 1 - 2\sin^2 x$

3. $\sin x\csc x = 1$
or: $\cos x\sec x = 1$
or: $\tan x\cot x = 1$

4. $\sin(-x) = -\sin x$
or: $\tan(-x) = -\tan x$
or: $\cot(-x) = -\cot x$
or: $\csc(-x) = -\csc x$

5. $\cot\theta = \tan(90° - \theta)$

6. $\cos 3x + \cos 7x = 2\cos 5x\cos 2x$

7. a. $\cos\dfrac{1}{2}\theta = -\dfrac{1}{\sqrt{6}}$
b. $\tan 2\theta = 4\sqrt{5}$

8. $\sin 15° = \dfrac{\sqrt{6} - \sqrt{2}}{4}$

9. $\dfrac{\sec G}{\sin G} - \dfrac{\sin G}{\cos G}$
$= \dfrac{1 - \sin^2 G}{\sin G\cos G}$
$= \dfrac{\cos^2 G}{\sin G\cos G}$
$= \cot G$
$\therefore \dfrac{\sec G}{\sin G} - \dfrac{\sin G}{\cos G}$
$= \cot G$

10. $S = \left\{-\pi, -\dfrac{\pi}{3}, \dfrac{\pi}{3}, \pi\right\}$

11. $S = \{-180°, -45°, 0°, 45°, 180°\}$

12. $S = \{47° + 90n°\}$

13. $y = 25\cos(\theta - 73.739...°)$
$\therefore$ high is at $\theta - 73.739...°$.

TEST 45 CHAPTER 15 Sections 15-1 to 15-6

1. $b = 69.3446... \approx 69.3$ in.

2. $d = 0.027544... \approx 0.028$ microns

3. $G = -8°$. No such triangle.

4. a. $\cos M = \dfrac{904}{952}$

$M = 18.2718... \approx 18.27°$

b. Area $= 74.619... \approx 74.62\ \text{cm}^2$

c. Expression $= 74.619... =$ Area.

5. $n = 1.05677... \approx 1.06$ miles

6. $R = 65.862... \approx 65.9°$

7. a. $u = 4.4865... \approx 4.5$ cm

b. (Reduced in size)

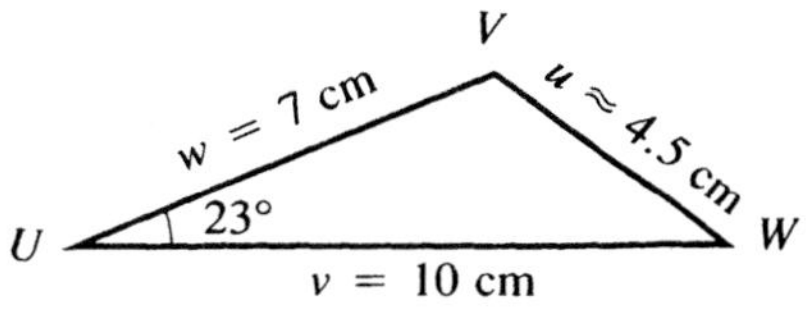

8. $\sin 2x = 2\sin x\cos x$

TEST 46 CHAPTER 15 Sections 15-7 and 15-8

1. $|\vec{r} + \vec{s}| = 10.975...,\ \theta = 104.52...°$

$|\vec{r} - \vec{s}| = 30.651...,\ \theta = 20.280...°$

2. a. 3.378... cm at 155.94...°

b.

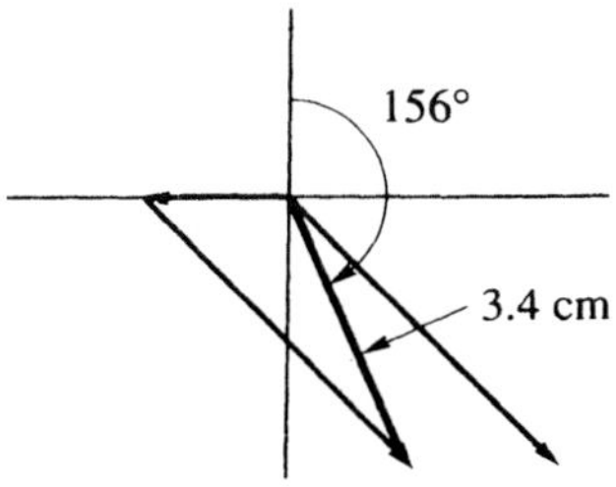

3. $-24.04...\vec{i} - 16.21...\vec{j}$

4. $5\vec{i} - 6\vec{j} = \sqrt{61}$ at 309.80...°

5. $S = \{-270°, 0°, 90°\}$ **6.** $\dfrac{24}{25}$

TEST 47 CHAPTER 15

1. a. 97°26′ b. 212.69... cm^2

2. $663.056 ... \approx 663$ ft

3. 247.861... at 9.321...°, or about 248 mph at 9.3°

4. Between 1.243... and 7.236..., or about 1.2 and 7.2 km

5. period = 26

TEST 48 CHAPTERS 13 through 15

1.

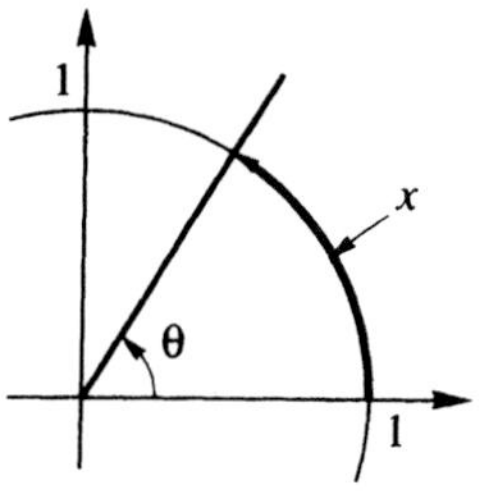

x for circular functions
θ for trig functions

2. $\sin\theta = \dfrac{v}{r},\ \sin x = v$ **3.** $\tan x = \dfrac{\sin x}{\cos x}$

4. $x = \cos^{-1} t$ if and only if $t = \cos x$

5. arcsec x is the *general* value.

Arcsec x is the *principal* value, in $[0, \pi]$, but not $\dfrac{\pi}{2}$.

6. a. $-\dfrac{2}{\sqrt{3}}$ b. $-\dfrac{1}{\sqrt{3}}$

c. -1 d. no value

e. $\dfrac{15}{17}$ f. $\dfrac{5\pi}{6}$

g. $\dfrac{\pi}{2}$

7. a.

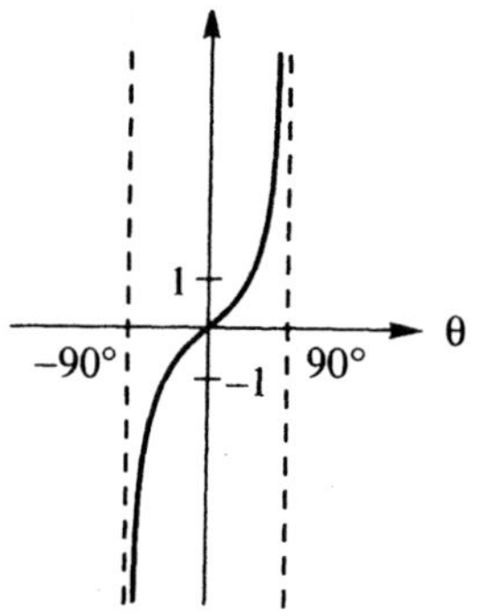

b.

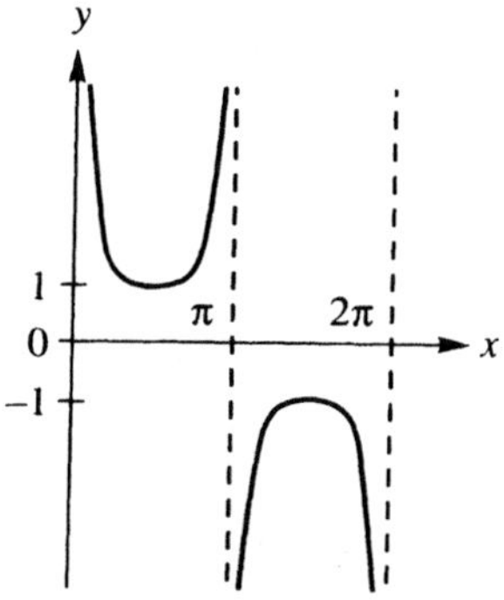

c.

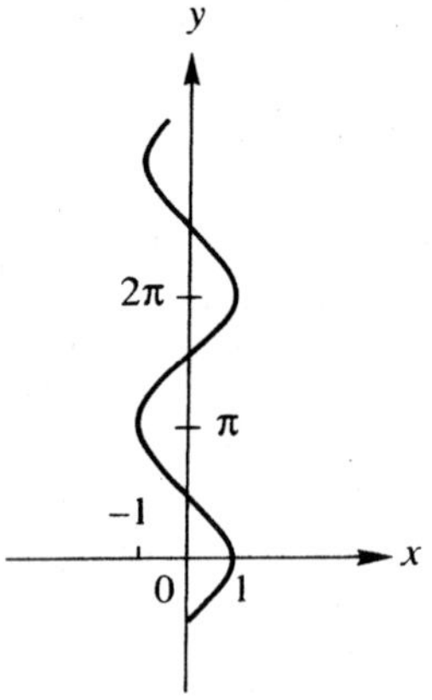

d.

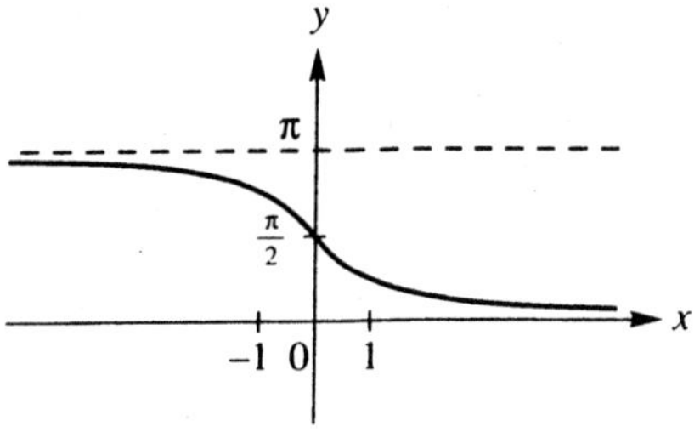

8. This range makes the graph continuous.

9.

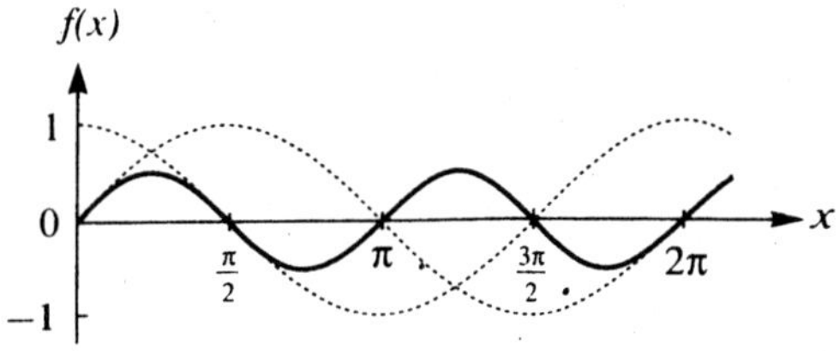

10. $y = 7 + 5\cos\frac{\pi}{6}(x - 3)$ or $y = 7 + 5\sin\frac{\pi}{6}x$

11. a. 10.4477...
b. 7.314..., 10.685..., 21.314...

12. a. $1 + \tan^2 x = \sec^2 x$
b. $\sec x = \dfrac{1}{\cos x}$
c. $\csc(-x) = -\csc x$
d. $\cos\left(\dfrac{\pi}{2} - x\right) = \sin x$
e. $\cos(A - B) = \cos A \cos B + \sin A \sin B$
f. $\sin 2x = 2 \sin x \cos x$
g. $\sin 5x \cos 7x = \frac{1}{2}\sin 12x - \frac{1}{2}\sin 2x$
h. $\cos 37° + \cos 95° = 2\cos 66° \cos 29°$
i. $5\sin\theta + 12\cos\theta = 13\cos(\theta - 22.6...°)$

13. $-\dfrac{1}{\sqrt{10}}$

14. $\sin x(\csc x - \sin x)$
$= 1 - \sin^2 x$
$= \cos^2 x$
$\therefore \sin x(\csc x - \sin x) = \cos^2 x$

15. $S = \{15°, 165°, 195°, 345°\}$

16. $\dfrac{\sin A}{a} = \dfrac{\sin B}{b} = \dfrac{\sin C}{c}$

17. a. 129.86...°
b. (Reduced in size)

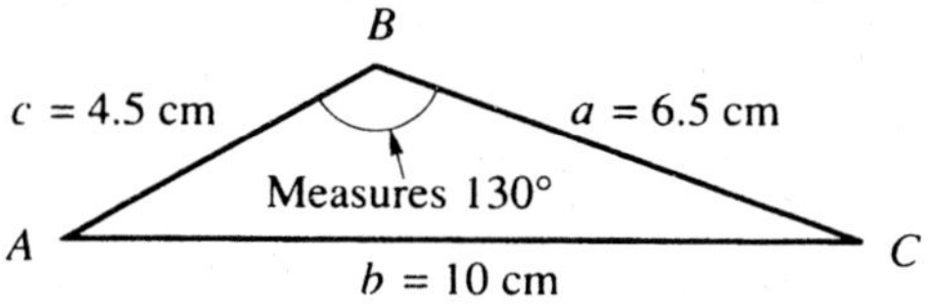

18. (Reduced in size)

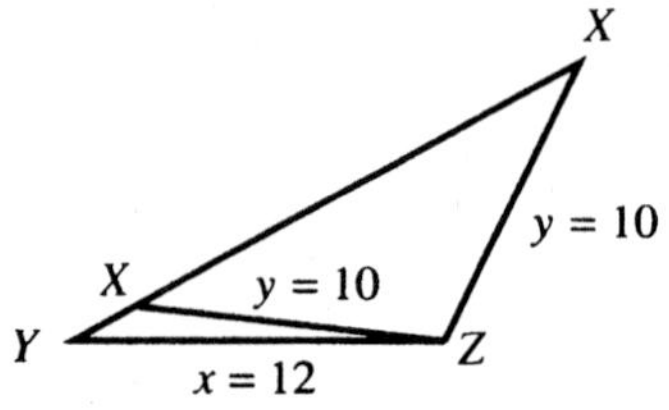

19. a. $-4\vec{i} + 6\vec{j}$
b. 7.211... at 123.69...°

TEST 49 FINAL EXAMINATION

1. a. $(xy)z = x(yz)$ b. Example: $x^2 + x + 1$
c. $|3x - 14| < 19$
$-19 < 3x - 14 < 19$
$-1\frac{2}{3} < x < 11$
$S = \{-1, 0, 1, \ldots, 10\}$

2. a. *Traffic Problem*

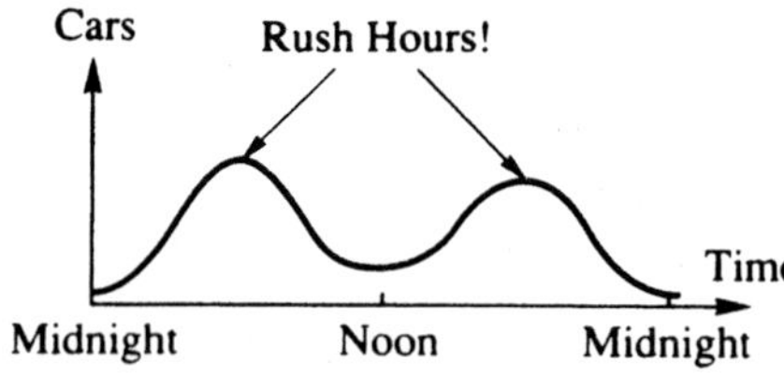

b. Example:

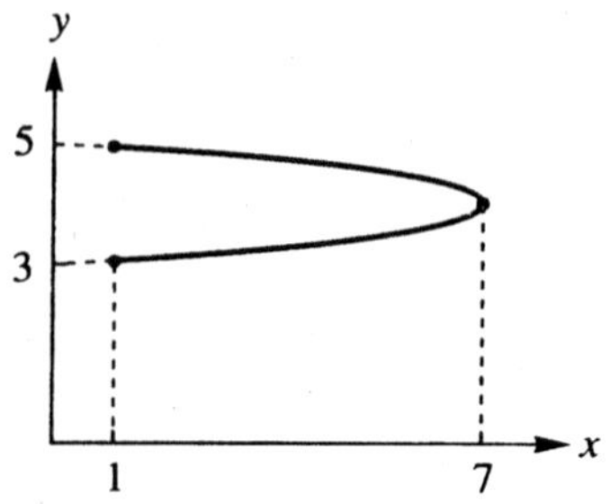

3. *Bookbinding Problem*

Let x = no. of pages.
Let y = no. of dollars.
(160, 6.60), (384, 11.08)

$$m = \frac{11.08 - 6.60}{384 - 160} = 0.02$$

$y - 6.60 = 0.02(x - 160)$

$\underline{y = 0.02x + 3.40}$

If $x = 800$, $y = 0.02(800) + 3.40$
$= \underline{\$19.40}$

4. *Tax Bracket Problem*

a. $f(x) = 330 + 0.15(x - 3000)$
$f(x) = 4080 + 0.28(x - 28{,}000)$
$f(x) = g(x)$:
$330 + 0.15x - 450 = 4080 + 0.28x - 7840$
$3650 = 0.13x$
$28{,}000 = x$

$\underline{\$28{,}000}$

b. $2x - y > 0 \Rightarrow y < 2x$
$x + y \geq 3 \Rightarrow y \geq -x + 3$
$y + 1 > \frac{1}{3}(x - 3)$

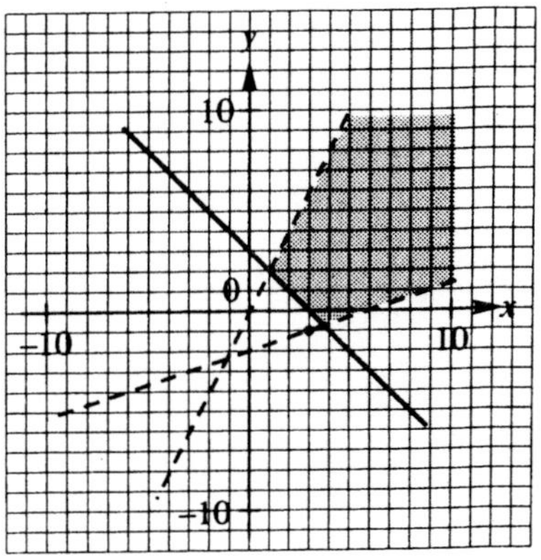

5. a. $y - k = a(x - h)^2$
(3, −4) vertex, (1, 0) other point.
$0 + 4 = a(1 - 3)^2$
$1 = a$
$\therefore y + 4 = (x - 3)^2$
$\underline{y = x^2 - 6x + 5}$

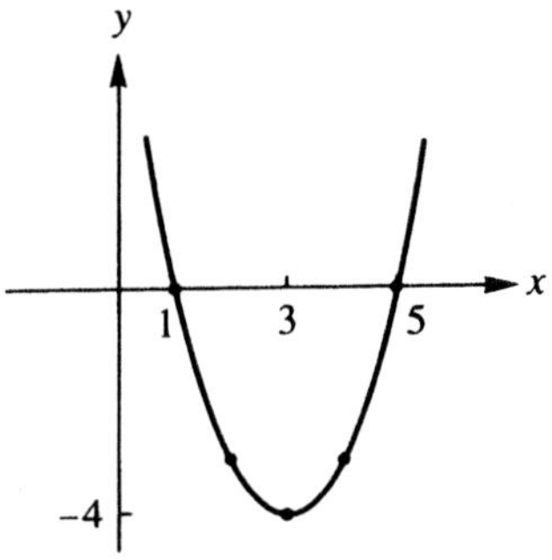

b. $x^2 - 8x + 41 = 0$

$$x = \frac{8 \pm \sqrt{-100}}{2}$$

$\underline{S = \{4 \pm 5i\}}$

6. $f(x) = a \times b^x$
(0, 200) and (5, 100)
$f(10) = 50$, $\underline{f(15) = 25}$
$C = 200$
$100 = 200 \times b^5$

$$b = \left(\frac{100}{200}\right)^{\frac{1}{5}} = 0.870550\ldots$$

$\therefore f(x) = \underline{200 \times (0.870550\ldots)^x}$

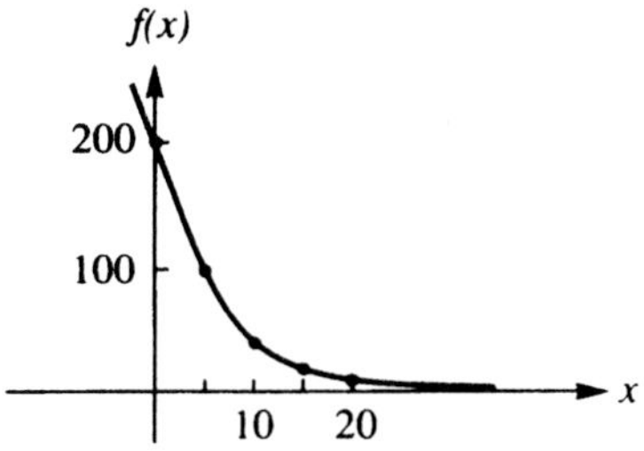

$f(x) = 3$:
$3 = 200 \times 0.870550\ldots^x$

$$x = \frac{\log\left(\frac{3}{200}\right)}{\log 0.870\ldots}$$

$\underline{x = 30.294\ldots}$

7. $$f(x) = \frac{6x^2 + 12x - 90}{x^3 + 4x^2 - 11x - 30}$$

$$f(x) = \frac{6(x + 5)(x - 3)}{(x + 2)(x - 3)(x + 5)}$$

$$f(x) = \frac{6}{x + 2},\ x \neq -2, 3, -5$$

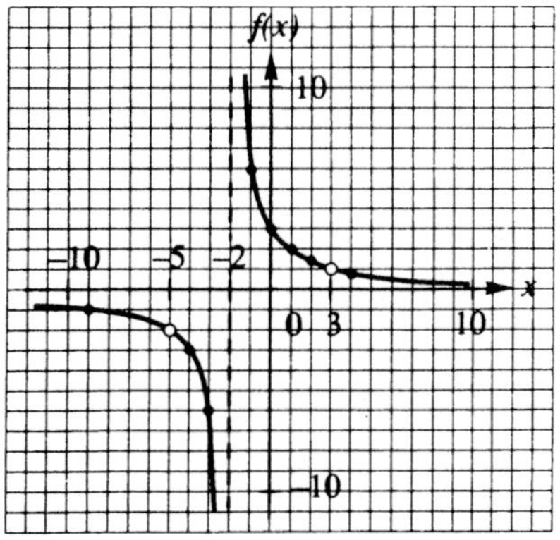

8. a. $$\frac{92}{\sqrt{3} + 7}$$
$$= \frac{92(\sqrt{3} - 7)}{3 - 49}$$
$$= \underline{-2\sqrt{3} + 14}$$

b. *Animal Body Problem*
Let m = no. of kg mass.
Let a = no. of cm^2 area.
$a = km^{\frac{2}{3}}$
$150 = k \cdot 80^{\frac{2}{3}}$
$k = 8.0791...$
$\therefore$ equation is $\underline{\underline{a = 8.0791...\, m^{\frac{2}{3}}}}$

$2000 = 8.0791...\, m^{\frac{2}{3}}$
$m = \left(\frac{2000}{8.0791...}\right)^{\frac{3}{2}}$
$m = 3894.91...$
About $\underline{\underline{\text{3900 grams}}}$

9. a. $25x^2 + 4y^2 + 150x - 32y + 189 < 0$
$25(x^2 + 6x + 9) + 4(y^2 - 8y + 15)$
$< -189 + 225 + 64$
$\left(\frac{x+3}{2}\right)^2 + \left(\frac{y-4}{5}\right)^2 < 1$

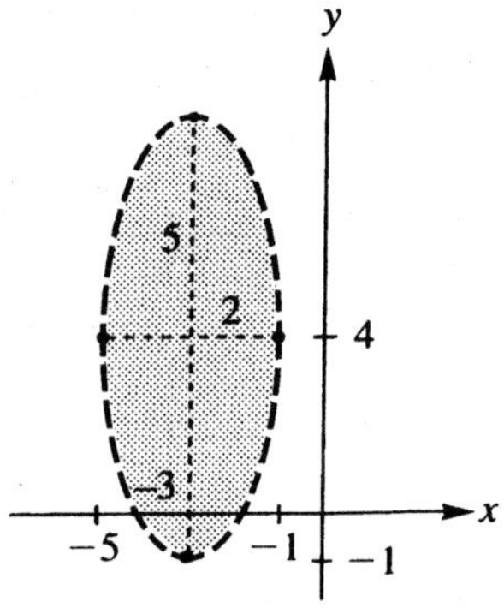

b. $x^2 + y^2 - 4x = 21$ ①
$x^2 + 5y^2 = 49$ ②
$5 \times$ ①: $5x^2 + 5y^2 - 20x = 105$
$x^2 + 5y^2 = 49$
$4x^2 - 20x = 56$

$x^2 - 5x - 14 = 0$
$(x - 7)(x + 2) = 0$
$x = 7$ or $x = -2$
$y = 0$ or $y = \pm 3$
$S = \{(7, 0), (-2, 3), (-2, -3)\}$

10. a. Example:

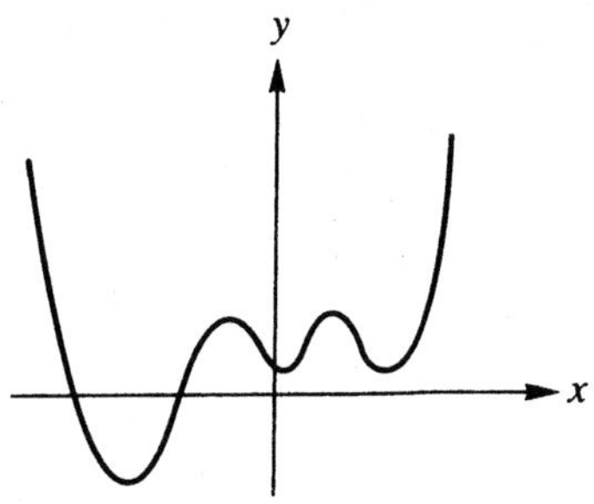

b. $f(x) = x^3 - 12x^2 + 46x - 52$

2⌋	1	−12	46	−52
		2	−20	52
	1	−10	26	0

$f(x) = (x - 2)(x^2 - 10x + 26)$
$f(x) = 0 \Leftrightarrow x = 2$ or $x^2 - 10x + 26 = 0$
Zeros are $\underline{\underline{2,\ 5 + i,\ 5 - i}}$

c. $f(x)$ has 3 sign reversals. So there could be 3 or 1 positive zero. Since $f(x)$ does have exactly 1 positive zero, Descartes' rule of signs is verified.

11. a. 3894, 3826, ...
$d = 3826 - 3894 = -68$
$t_{78} = 3894 + 77(-68)$
$= \underline{\underline{-1342}}$

b. $S_{60} = 300 \cdot \frac{1 - 1.03^{60}}{1 - 1.03}$
$= \underline{\underline{48916.03...}}$

c. $100 + 90 + 81 + \ldots\ r = 0.9$
$S = \frac{100}{1 - 0.9}$
$= \underline{\underline{1000}}$

d. $(a - b)^{11}$
b^5 term: $-\frac{11!}{6!5!}a^6b^5$
$= \underline{\underline{-462a^6b^5}}$

12. a. ${}_7P_3 = 7 \cdot 6 \cdot 5 = \underline{\underline{210}}$

b. $P(\text{Face}) = \frac{12}{52} = \underline{\underline{\frac{3}{13}}}$

c. $P(3, 6\text{'s}) = {}_5C_3 \cdot \left(\frac{1}{6}\right)^3\left(\frac{5}{6}\right)^2$
$= \frac{250}{7776}$
$= \frac{125}{3888}$
$\approx \underline{\underline{3.2\%}}$

b.

x	dev	dev^2
37	−1	1
35	−3	9
40	2	4
38	0	0
40	2	4
190		18

Mean $= \frac{190}{5} = \underline{\underline{38}}$

Std. dev. $= \sqrt{\frac{18}{(5 - 1)}}$
$= \underline{\underline{2.121...}}$

13. a. period $= 2(45 - 11) = 68$
$C = \frac{1}{2}(25 + 5) = 15$
$A = 25 - 15 = 10$
$\therefore$ equation is $\underline{\underline{y = 15 + 10 \cos \frac{\pi}{34}(x - 45)}}$

$y = 5.3$ when $x \approx 8.34,\ 13.66,\ 76.34$

b. $y = \sec\theta$

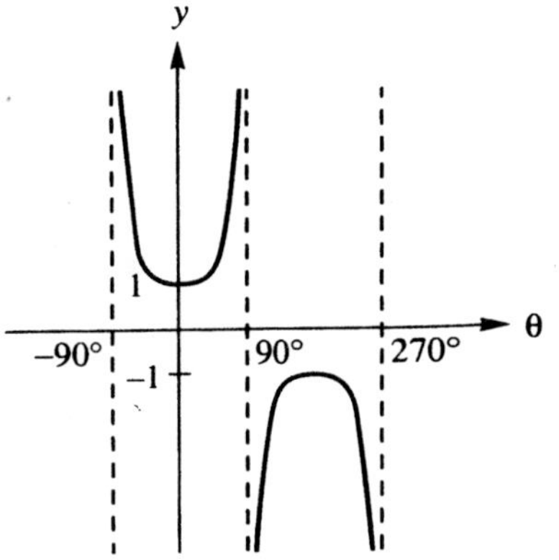

14. a. $1 + \tan^2 x = \sec^2 x$

b. $\cos^2 x - \sin^2 x = 0.5, \quad x \in (-\pi, \pi)$

$\cos 2x = 0.5$

$2x = \pm\frac{\pi}{3} + 2\pi n$

$x = \pm\frac{\pi}{6} + \pi n$

$S = \left\{\frac{\pi}{6}, -\frac{\pi}{6}, \frac{5\pi}{6}, -\frac{5\pi}{6}\right\}$

15. a.

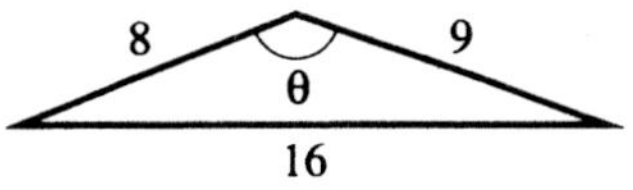

$\cos\theta = \frac{8^2 + 9^2 - 16^2}{2 \cdot 8 \cdot 9}$

$= -\frac{111}{144}$

$\theta = \underline{\underline{140.428\ldots^\circ}}$

$\text{Area} = \frac{1}{2} \cdot 8 \cdot 9 \cdot \sin\theta = \underline{\underline{22.933\ldots}}$

b.

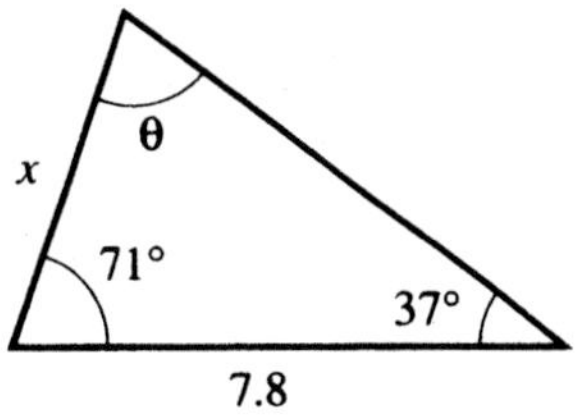

$\theta = 180° - (71° + 37°) = 72°$

$x = \frac{7.8 \sin 37°}{\sin 72°}$

$x = 4.9357\ldots$

$\approx \underline{\underline{4.9 \text{ cm}}}$